Chemtrails: The Silent Spread, Unraveling the Menace of Airborne Chemicals

Frederic Spinola

DEDICATION

To all those who have tirelessly pursued knowledge, questioned the unknown, and sought to uncover the truth, this book is dedicated to you. May it serve as a beacon of awareness, sparking curiosity and inspiring critical thinking about the controversial topic of chemtrails. Together, let us delve into the intricacies of this phenomenon, exploring its history, scientific perspectives, and societal implications. May our collective efforts foster a deeper understanding and open dialogue, ultimately leading us closer to the truth.

Índice

ACKNOWLEDGMENTS

May this book serve as a testament to the collaborative efforts of all those involved, as we strive to better comprehend the complex and multifaceted subject of chemtrails.

Lastly, we express our sincere appreciation to the readers who have embarked on this journey with us. Your curiosity and open-mindedness signify the importance of seeking knowledge and fostering informed discussions.

Our gratitude extends to our friends, family, and loved ones for their unwavering encouragement and understanding throughout the process of writing and researching this book. Your constant support has been a source of inspiration and motivation.

Furthermore, we acknowledge the support and guidance of our editors, proofreaders, and publishing team, whose meticulous efforts ensured the accuracy and readability of this book. Your commitment to excellence is greatly appreciated.

We would also like to thank the individuals who generously shared their personal experiences, anecdotes, and observations related to chemtrails. Your contributions have added a human touch and personal perspective, enriching the narrative within these pages.

Firstly, we extend our deepest appreciation to the researchers, scientists, and experts in relevant fields whose valuable insights and extensive knowledge have helped shape the content of this book. Your dedication to unraveling the mysteries surrounding chemtrails has been instrumental in providing a comprehensive understanding.

We express our heartfelt gratitude to all those who have contributed to the creation of this book, dedicating their time, expertise, and support to shed light on the subject of chemtrails.

CHEMTRAILS

The Silent Spread
Unraveling the Menace of Airborne Chemicals

ALL THE FACTS BY
FREDERIC SPINOLA

Chapter 1: The Invisible Enemy

As human beings, we have become accustomed to relying on our senses to navigate the world around us. Sight, sound, touch, taste, and smell guide us in our daily lives, helping us make sense of our environment. Yet, there is an enemy lurking in our midst that evades our senses entirely - airborne chemicals.

The air we breathe, seemingly innocuous, carries with it a multitude of substances that can either enrich or harm us. While many of these chemicals are harmless or even beneficial, there are those that pose a significant threat to our health and well-being. The insidious nature of these invisible foes lies in their ability to infiltrate every aspect of our lives without detection, silently spreading their menace.

Consider the everyday activities we engage in without a second thought. We wake up in the morning, prepare a cup of coffee, and toast a slice of bread. As we savor the aroma of freshly brewed coffee, we remain oblivious to the potential dangers that may be lurking in the air. Harmful chemicals can be released from the material components of our appliances, finding their way into the air we breathe and eventually into our bodies.

Indoor environments provide fertile ground for airborne chemicals to accumulate. Our homes, workplaces, and public spaces contain countless sources of potential pollutants, ranging from cleaning products to building materials. Many of these substances emit volatile organic compounds (VOCs), which can cause a range of health problems, including respiratory irritation, allergies, and even long-term damage to our organs.

The problem doesn't end with indoor environments. Outdoor air pollution has become a global concern, with urban areas experiencing alarmingly high levels of toxic chemicals. Vehicle emissions, industrial facilities, and even natural phenomena contribute to the presence of hazardous airborne pollutants. The consequences of this silent spread are far-reaching, affecting not only our physical health but also the quality of our environment and the well-being of future generations.

Understanding the risks posed by airborne chemicals is crucial for our society's progress and the safeguarding of public health. It is our responsibility to delve into the depths of this complex issue and shed light on its hidden dangers. This book aims to unravel the enigma of airborne chemicals, presenting the latest scientific research and case studies that illustrate their impact on human health and the environment.

In the chapters that follow, we will explore the various sources of airborne chemicals and their potential effects on different aspects of our lives. We will delve into the intricacies of indoor air quality, examining the common culprits that contribute to its degradation. From the chemicals present in cleaning products to those released by furniture and building materials, we will expose the silent spread of these insidious substances.

Additionally, we will venture into the great outdoors and investigate the far-reaching consequences of outdoor air pollution. We will examine the role of industrial activities, transportation, and natural factors in the pollution of our skies. Moreover, we will delve into the global efforts being made to mitigate these risks and find viable solutions for a cleaner, healthier world.

As you embark on this journey through the pages of "The Silent Spread: Unraveling the Menace of Airborne Chemicals," remember that awareness is the first step towards change. By arming ourselves with knowledge, we can take action to protect ourselves, our loved ones, and the environment we call home. The invisible enemy may be formidable, but together, we can conquer its silent spread.

The Invisible Enemy:

As we navigate the complexities of the modern world, it is of utmost importance to recognize the pervasive presence of airborne chemicals and their potential risks. To fully comprehend the impacts of these invisible enemies, we must delve deep into the scientific realm and uncover the intricacies of their spread.

Research has revealed that indoor air quality plays a critical role in our overall well-being. With the average person spending about 90% of their time indoors, whether at home, work, or in public spaces, it is crucial to assess the factors that contribute to the degradation of the air we breathe.

One significant source of indoor air pollution is the various household cleaning products we utilize on a daily basis. While they promise cleanliness and freshness, many of these products contain chemicals that can release harmful pollutants into the air. VOCs, such as formaldehyde and benzene, are commonly found in cleaning agents and can lead to headaches, respiratory irritation, and even more severe health issues with prolonged exposure.

Furthermore, the furniture we choose to furnish our homes and workplaces can also release harmful chemicals into the air. Some upholstery fabrics, carpets, and synthetic materials may contain flame retardants or polychlorinated biphenyls (PCBs), which have been linked to hormonal disruptions and neurological problems. Even seemingly innocuous items like scented candles or air fresheners can release volatile compounds, contributing to a decline in indoor air quality.

Beyond our indoor spaces, the silent spread of airborne chemicals extends to the great outdoors, where a multitude of factors contribute to the deterioration of outdoor air quality. Urban areas, in particular, suffer from high levels of toxic pollutants due to the concentration of human activities and the presence of industrial facilities.

Vehicle emissions remain a significant contributor to outdoor air pollution, with exhaust fumes releasing a cocktail of chemicals into the atmosphere. Fine particulate matter, nitrogen oxides, and volatile organic compounds are among the pollutants that contribute to the formation of smog and pose a threat to human health. Respiratory illnesses, cardiovascular problems, and even premature death can be attributed to long-term exposure to these pollutants.

Industrial activities, such as manufacturing processes and power generation, also release copious amounts of airborne chemicals. Industrial emissions comprise a range of hazardous substances, including heavy metals, sulfur dioxide, and ozone-depleting agents. These pollutants not only harm human health but also have adverse effects on the environment, contributing to climate change and biodiversity loss.

Natural phenomena, such as wildfires and volcanic eruptions, also play a role in the global spread of airborne chemicals. While these events are beyond human control, their impact on air quality cannot be ignored. Smoke and ash from wildfires can travel vast distances, affecting even remote areas, and volcanic eruptions can release large quantities of sulfur dioxide, which can cause acid rain and widespread respiratory issues.

Amidst the silent spread of airborne chemicals, it is essential to recognize the collective responsibility we bear in safeguarding public health and the environment. Governments, regulatory bodies, and scientific communities worldwide are working tirelessly to develop strategies to mitigate the risks associated with these invisible foes.

Stringent regulations on emissions, the promotion of clean energy sources, and the development of sustainable transportation alternatives are some of the measures being implemented in many countries. Awareness campaigns and educational initiatives also play a pivotal role in empowering individuals to make informed decisions regarding their daily activities.

As we conclude this exploration of airborne chemicals and their silent spread, it is crucial to reiterate the power of knowledge and awareness. By arming ourselves with scientific research, understanding the sources of airborne chemicals, and recognizing their potential effects, we can take proactive steps towards reducing our exposure and protecting the well-being of ourselves and future generations.

Together, let us face the invisible enemy head-on, driven by a shared determination to create a cleaner, healthier world. The silent spread of airborne chemicals may be formidable, but with knowledge and action, we can conquer this menace and secure a brighter future for all.

Chemtrails

Chapter 2: A Brief History of Airborne Chemicals

Tracing back to ancient times, this chapter delves into the historical use and exposure to airborne chemicals and how it has evolved over centuries.

Human history is dotted with various discoveries and inventions that have shaped the course of civilization. Among these developments, the harnessing and understanding of airborne chemicals have played a significant role in both beneficial and detrimental ways. From the earliest civilizations to the modern age, the human race has experienced an intricate relationship with airborne chemicals, often inadvertently exposing themselves to their potential dangers.

In the annals of ancient civilizations, the use of airborne chemicals was prevalent, though not always recognized as such. The Egyptians, for instance, extensively employed incense, oils, and resins in their religious ceremonies and rituals. These aromatic substances were believed to communicate with the gods, purify the air, and ward off evil spirits. Little did they know that the smoke emitted during their ceremonies contained airborne chemicals that could have detrimental health effects with prolonged exposure.

Moving forward in time, the influence of airborne chemicals remained pervasive. In ancient China, scholars and alchemists sought to unlock the secrets of immortality through experimental elixirs. These concoctions were often derived from various natural compounds, herbs, and minerals. Sadly, these ambitious pursuits sometimes resulted in unintended exposure to toxic fumes and vapors. Although the true nature of these adverse effects may not have been explicitly understood, the connection between exposure to certain airborne chemicals and their potential harm was established, even in ancient times.

With the rise of the medieval period, advancements in chemistry, alchemy, and medicine laid the foundation for a more intricate understanding of airborne chemicals. In Europe, alchemists began distilling various substances to obtain valuable materials and explore their properties. Paracelsus, a renowned Swiss physician and alchemist of the 16th century, believed that both the dose and the specific airborne chemical determined whether it was a poison or a remedy. This revolutionary concept laid the groundwork for future investigations into the effects of airborne chemicals on human health.

During the scientific revolution of the 17th and 18th centuries, researchers like Robert Boyle and Antoine Lavoisier made significant contributions towards understanding the properties of airborne chemicals. These pioneers developed more sophisticated methods of analysis and measurement, laying the groundwork for quantitative investigations. As the scientific community's knowledge expanded, the understanding of the composition and behavior of airborne chemicals grew increasingly nuanced.

The industrial revolution, with its rapid advancements in technology and manufacturing, brought about both incredible progress and unforeseen consequences. Chemicals were increasingly harnessed for diverse purposes, such as in the production of textiles, dyes, and fuels. However, the unregulated use and exposure to these chemicals in workplaces and communities led to severe health ailments. As production escalated to meet the demands of a growing population, the detrimental effects of exposure to hazardous airborne substances became undeniable.

In the early 20th century, the world witnessed an alarming incident that underscored the dangers of airborne chemicals like never before. The Bhopal disaster, which occurred in 1984, remains one of the most catastrophic industrial accidents in history. A release of methyl isocyanate gas from a pesticide plant in Bhopal, India, led to the immediate deaths of thousands and the long-term suffering of countless others. This horrifying event served as a stark reminder of the potential devastation caused by the mishandling and release of airborne chemicals.

As our understanding of airborne chemicals continues to evolve, it becomes imperative to navigate a delicate balance between harnessing their benefits and mitigating their risks. In the modern era, industries have made significant strides in implementing safety measures and regulations to protect workers and the environment. However, new challenges continually emerge with the introduction of novel airborne chemicals and their varying impacts on human health.

The story of airborne chemicals is far from over. In the next half of this chapter, we will explore the more recent advancements, the increasing prevalence of airborne pollutants, and the efforts made towards mitigating their detrimental effects. So, join us as we delve deeper into this fascinating and critical aspect of our interconnected world.

The understanding and exploration of airborne chemicals have come a long way since ancient times. As we delve deeper into this chapter on the history of airborne chemicals, it becomes evident that the human race has continuously grappled with the complex relationship between these substances and their potential harm. From the scientific revolution to the modern era, advancements in chemistry, technology, and medicine have provided us with a more nuanced understanding of airborne chemicals and their impacts on our health and the environment.

In the late 19th century, the field of toxicology emerged as scientists sought to understand the effects of chemicals on the human body. This multidisciplinary branch of science delved into the study of airborne chemicals and their potential toxicity, further deepening our knowledge on the subject. During this time, the detrimental effects of toxic substances such as lead and mercury became increasingly apparent. From the tragic cases of occupational exposures in industries like hat-making, where the use of mercury in the production of felt led to the development of neurological disorders among workers, to the widespread lead poisoning prevalent in communities due to the use of lead-based products like paint and gasoline, the harmful effects of airborne chemicals were undeniable.

The 20th century marked a turning point in our ability to measure, study, and regulate airborne chemicals. Rapid industrialization and technological advancements brought about new challenges as well as opportunities for progress. Chemical manufacturing escalated to meet societal demands, and with it came the release of numerous synthetic compounds into the air. Many of these chemicals were beneficial and revolutionized daily life, such as antibiotics and plastics. However, concerns began to arise regarding their potential adverse effects on human health and the environment.

The increased understanding of the impacts of airborne chemicals on both human and environmental health led to the establishment of regulatory frameworks to safeguard against their potential harm. In the United States, for example, the 1970 Clean Air Act (CAA) was enacted to regulate air pollution, including the emissions of hazardous airborne chemicals from industrial processes and transportation. The CAA empowered the Environmental Protection Agency (EPA) to set national air quality standards, establish limits for emissions, and enforce regulations to protect public health.

Despite these regulatory efforts, new challenges continue to emerge in the modern era. With rapid industrialization in developing countries and the introduction of novel chemicals and pollutants, concerns about air quality and the health impacts of airborne chemicals persist. As human activities expand, so does the potential for unintended exposure to harmful substances. For instance, the burning of fossil fuels, industrial emissions, and the use of certain consumer products release volatile organic compounds (VOCs) into the atmosphere. These VOCs can react with other pollutants and sunlight, leading to the formation of ground-level ozone, a harmful air pollutant that contributes to respiratory issues and environmental damage.

Furthermore, the increasing prevalence of airborne pollutants has sparked a growing interest in the field of indoor air quality. In recent decades, studies have highlighted the potential risks associated with indoor environments, where people spend a significant portion of their time. Indoor air can be contaminated with various chemicals from sources such as building materials, furnishings, cleaning products, and even cooking activities. Understanding and managing these indoor pollutants have become essential for maintaining healthy living and working environments.

In response to these emerging challenges, researchers, policymakers, and industries are actively working towards identifying and mitigating the detrimental effects of airborne chemicals. The development of sophisticated monitoring technologies allows for more accurate measurements of air quality and real-time data collection. Additionally, advancements in analytical chemistry enable the identification and quantification of specific airborne chemicals, helping us understand their sources and potential health risks.

Moreover, public awareness campaigns have emphasized the importance of individual actions in reducing exposure to harmful airborne chemicals. Promoting proper ventilation, using natural cleaning products, and monitoring indoor air quality are just a few examples of steps individuals can take to mitigate the risks associated with airborne pollutants.

In conclusion, the history of airborne chemicals is a complex and fascinating story that spans the centuries. From early civilizations unknowingly exposing themselves to potential hazards in their religious rituals to the modern era's regulations and advancements in science, our understanding of airborne chemicals and their impacts has greatly evolved. While progress has been made in mitigating the detrimental effects of these substances, challenges persist in the form of emerging pollutants and increasing industrialization. As we continue to navigate this intricate world of airborne chemicals, it is essential to strike a balance between harnessing their benefits and safeguarding human health and the environment.

Chapter 3: The Science Behind Airborne Chemicals

Exploring the mechanisms and properties of airborne chemicals, this chapter provides a scientific foundation for understanding their behavior in the environment.

Airborne chemicals have long been a subject of intrigue and concern. These invisible agents, suspended in the air we breathe, have the potential to exert a profound impact on our health and the environment. As we delve into the science behind airborne chemicals, we uncover a complex web of interactions and properties that shape their behavior and effects.

At the heart of understanding airborne chemicals lies the field of atmospheric chemistry, which explores the composition and transformations of substances in the Earth's atmosphere. Unlike liquids or solids, airborne chemicals are typically present as gases or aerosols, enabling them to disperse over vast distances with ease. This ability to travel through the air contributes to their potential for widespread impact.

One key aspect to consider when examining airborne chemicals is their sources. These chemical compounds can originate from both natural and anthropogenic activities. Natural sources include emissions from vegetation, oceans, and volcanic activity, releasing substances such as volatile organic compounds (VOCs) and sulfur dioxide (SO2) into the air. On the other hand, human activities contribute significantly to the release of airborne chemicals, ranging from industrial processes and transportation to everyday household activities like cooking and cleaning.

Once these chemicals are released into the atmosphere, their fate is determined by a complex set of factors. Physical and chemical properties such as volatility, solubility, and reactivity play a crucial role in shaping their behavior and distribution. Volatile substances, for instance, readily evaporate into the air, making them more likely to be inhaled or dispersed over larger areas. Conversely, less volatile compounds may persist in the atmosphere for longer periods, potentially undergoing chemical reactions or being deposited onto surfaces.

The role of temperature and atmospheric conditions cannot be overlooked when studying the behavior of airborne chemicals. Warmer temperatures tend to increase the volatility of substances, promoting their evaporation into the air. Atmospheric stability, turbulence, and wind patterns also influence the transport and dispersal of these chemicals, determining the areas where they may accumulate or be diluted.

Furthermore, the behavior of airborne chemicals is intricately linked to our understanding of air pollution and its impacts. Particulate matter, a type of airborne chemical consisting of tiny solid or liquid particles suspended in the air, has been a particular focus of research due to its detrimental effects on human health. These particles can be directly emitted or formed through chemical reactions in the atmosphere, contributing to respiratory and cardiovascular diseases.

In recent years, the study of airborne chemicals has gained added significance due to the emerging understanding of aerosols' role in disease transmission. The ongoing COVID-19 pandemic has highlighted the potential for certain pathogens, including the SARS-CoV-2 virus, to be transported through aerosols. Understanding the behavior of airborne chemicals and their interaction with pathogens is crucial for implementing effective mitigation strategies and safeguarding public health.

As we begin to unravel the intricate web of mechanisms and properties that govern the fate of airborne chemicals, questions arise. How do different substances interact with one another in the atmosphere? How do these chemicals transform under different environmental conditions? What are the long-term implications of their presence in the air we breathe?

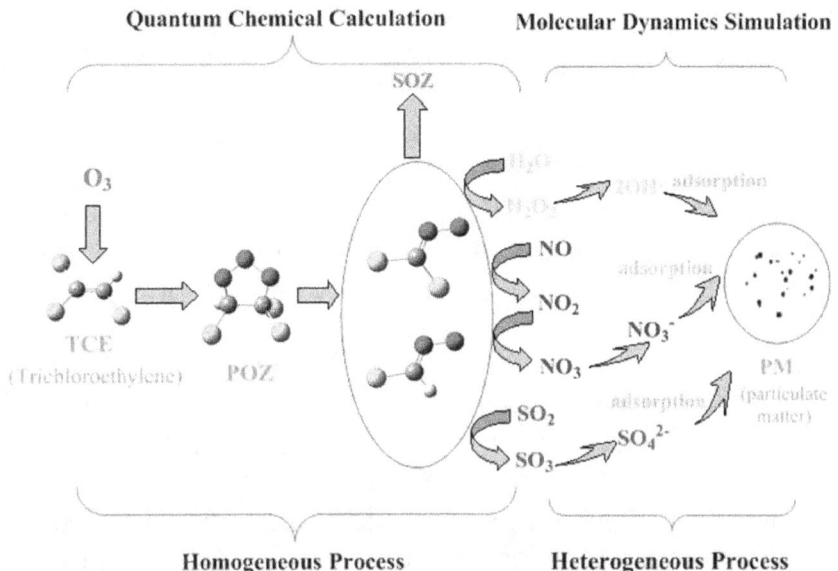

In the next half of this chapter, we will delve deeper into these questions, exploring the intricate processes involved in the dispersion and fate of airborne chemicals. By understanding the dynamic nature of these invisible agents, we can devise strategies to mitigate their potential risks and protect both human and environmental well-being.

In understanding the intricate processes involved in the dispersion and fate of airborne chemicals, it becomes evident that the interactions between different substances within the atmosphere play a crucial role. Chemical reactions occurring in the atmosphere can lead to the formation of new compounds, altering the composition and properties of airborne chemicals.

One important mechanism that affects the behavior of airborne chemicals is oxidation. Many compounds present in the atmosphere, such as volatile organic compounds (VOCs) and nitrogen oxides (NOx), undergo reactions with atmospheric oxidants, primarily hydroxyl radicals (OH). These reactions, known as oxidation, can result in the formation of secondary organic aerosols (SOAs) and other reaction products. SOAs are a significant component of particulate matter in the atmosphere and have the potential to influence air quality and human health.

The formation and transformation of SOAs involve complex chemical pathways and depend on various factors such as temperature, humidity, the abundance of precursor compounds, and the availability of atmospheric oxidants. Laboratory and field studies have demonstrated that SOAs can be formed through both gas-phase and particle-phase reactions. Understanding these processes is essential for accurately predicting the composition and behavior of airborne chemicals in different atmospheric conditions.

In addition to chemical reactions, physical processes such as condensation and evaporation play a vital role in the behavior of airborne chemicals. Volatile substances, including certain organic compounds, can exist as vapors in the atmosphere and undergo rapid evaporation and condensation processes. This volatility affects their distribution, as more volatile compounds tend to be present in the gas phase, while less volatile substances are more likely to be associated with aerosols or particles.

The condensation of vapor molecules onto existing particles can lead to the growth of aerosols through a process called nucleation. Nucleation occurs when gaseous molecules cluster together to form small particles that can grow in size by condensing additional vapor molecules. These newly formed particles may act as cloud condensation nuclei (CCN), influencing cloud formation and properties.

Understanding the behavior of airborne chemicals necessitates considering their fate and deposition onto various surfaces. Deposition processes depend on particle size, composition, and characteristics of the surfaces encountered. Larger particles, such as coarse particulate matter, tend to deposit more quickly due to gravitational settling, while smaller particles are subject to other mechanisms such as diffusion and interception. The deposition of particles onto surfaces can have environmental implications, influencing soil and water quality and potentially serving as a source of exposure for organisms.

Moreover, once deposited, airborne chemicals can either persist or undergo further transformations. Chemical reactions facilitated by sunlight, known as photoreactions, can occur on surfaces and lead to the formation of new compounds. These reactions have been observed on building surfaces, vegetation, and airborne particles. Photoreactions can result in the formation of reactive oxygen species and secondary pollutants, further complicating the understanding of the fate and behavior of airborne chemicals.

The long-term implications of airborne chemicals are of great concern for both human and environmental health. Exposure to elevated levels of certain airborne chemicals, such as particulate matter, has been linked to respiratory and cardiovascular diseases, as well as adverse impacts on ecosystems. Additionally, the potential for certain pathogens to be transmitted through airborne particles, as evidenced by the COVID-19 pandemic, highlights the need for understanding the behavior of airborne chemicals in disease transmission and implementing effective mitigation strategies.

By unraveling the intricate web of mechanisms and processes involved in the dispersion and fate of airborne chemicals, scientists and policymakers can develop strategies to mitigate potential risks and protect human and environmental well-being. Through continued research, monitoring, and regulation, it is possible to better understand and address the menace of airborne chemicals, ensuring a healthier and safer future for us all.

Chapter 4: Common Sources of Airborne Chemicals

Airborne chemicals are a ubiquitous presence in our lives, often unnoticed yet silently infiltrating our surroundings. While some of these chemicals may be harmless, others can pose serious risks to our health and the environment. In this chapter, we delve into the diverse sources that contribute to the presence of airborne chemicals, shedding light on everyday objects and activities that emit them.

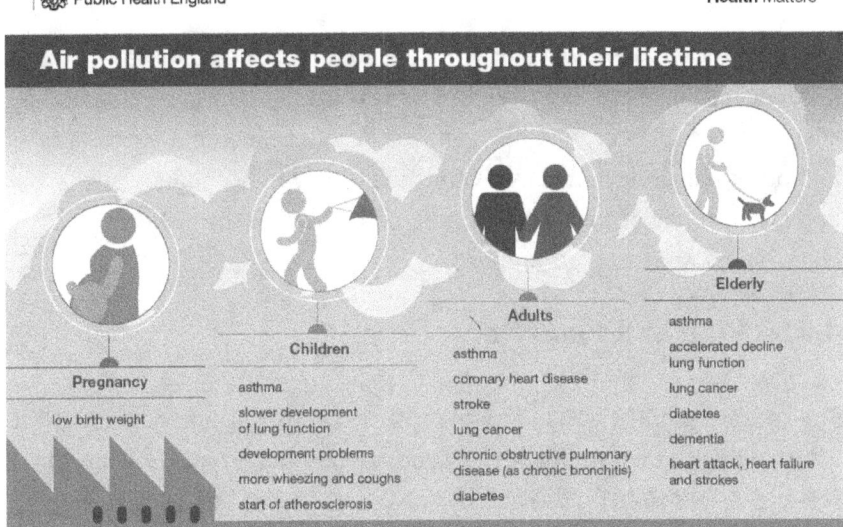

1. Household Cleaning Products:

Walk into any supermarket, and you'll be greeted by rows of cleaning products promising a spotless home. However, within those attractive bottles, lie hidden dangers. Cleaning products such as sprays, aerosols, and even seemingly innocent air fresheners often contain a cocktail of volatile organic compounds (VOCs). These VOCs can contribute to indoor air pollution, which can be detrimental to our respiratory systems when inhaled.

2. Building Materials and Furnishings:

The very structure that provides us shelter can also emit harmful airborne chemicals. Many building materials, such as paint, adhesives, carpets, and composite wood products, release VOCs into the air. These chemicals off-gas over time and can cause headaches, dizziness, and even respiratory problems for those constantly exposed to them. Additionally, flame retardants used in furniture and textiles can release toxic fumes when exposed to fire, posing a potential hazard.

3. Personal Care Products:

From the moment we wake up until we go to bed, personal care products are an integral part of our daily routines. However, certain cosmetics, fragrances, and hair care products may contain phthalates and other harmful chemicals. Even though they may enhance our appearances, these products can also contribute to the airborne chemical composition around us.

4. Cooking and Food Preparation:

The smell of a home-cooked meal is often comforting and nostalgic, but it's essential to be mindful of the airborne chemicals that cooking can release. As we whip up delicious dishes, the oils, fats, and spices we use can produce smoke and vapors containing particles and chemicals. These include polycyclic aromatic hydrocarbons (PAHs) and acrolein, which can have harmful effects on our health.

5. Tobacco Smoke:

The dangers of tobacco smoke are well-established, but the impact extends beyond the immediate smoker. Secondhand smoke, filled with harmful particulates and chemicals, can linger in the air for hours, endangering the health of those around. Even thirdhand smoke, the residue left behind on surfaces after smoking has ceased, can emit toxic compounds into the air, increasing the risk of respiratory issues and other health problems.

6. Household Pesticides:

To keep our homes free from pests, many of us turn to pesticides. However, the same chemicals designed to eliminate unwanted intruders can also become airborne, posing a threat to both human and animal health. Inhaling pesticides can lead to symptoms such as headaches, dizziness, nausea, and in severe cases, even organ damage.

7. Electronic Devices and Indoor Air Pollution:

In our modern world, electronic devices have become indispensable. However, what often goes unnoticed is the release of airborne chemicals from these devices. Computers, printers, and even the paint on our walls can emit low levels of chemicals such as formaldehyde and benzene, contributing to indoor air pollution, and potentially causing respiratory irritation and other health issues.

As we have discovered, the sources of airborne chemicals are plentiful and diverse. From the cleaning products we use to the very materials that make up our homes, these substances surround us at every turn. However, it is crucial to remember that awareness is the first step towards reducing our exposure and mitigating potential risks. In the second half of this chapter, we will explore strategies to minimize airborne chemical exposure and create healthier living spaces. Stay tuned for the next installment as we uncover effective measures to combat the menace of airborne chemicals.

In the second half of this chapter, we will explore strategies to minimize airborne chemical exposure and create healthier living spaces. By implementing these measures, we can take proactive steps to safeguard our health and well-being in an environment infiltrated by potentially harmful substances.

8. Ventilation and Air Filtration:

Good ventilation is paramount in reducing the concentration of airborne chemicals indoors. Opening windows and doors to allow fresh air to circulate can help dilute pollutants. Additionally, using air filters or purifiers can trap and remove harmful particles from the air, improving indoor air quality. It is particularly important to ensure proper ventilation in areas where chemical-intensive activities, such as painting or cooking, take place.

9. Green Cleaning Products:

As we become more aware of the risks associated with conventional cleaning products, the demand for eco-friendly alternatives has surged. Many manufacturers now offer cleaning products that are free from harmful chemicals, such as VOCs and phthalates. By opting for these green alternatives, we can maintain a clean home without compromising our health.

10. Proper Storage and Disposal:

The storage and disposal of chemicals are often overlooked aspects when it comes to mitigating their impact on indoor air quality. Ensuring that chemicals, including cleaning agents and pesticides, are stored in well-ventilated areas can minimize the risk of off-gassing. Additionally, following proper disposal protocols for unused or expired chemicals can prevent them from infiltrating the environment and contributing to pollution.

11. Indoor Plants:

Nature has a remarkable ability to improve air quality, and indoor plants are no exception. Certain plants, such as peace lilies, spider plants, and bamboo palms, have been found to help remove toxins from the air. These plants absorb airborne chemicals through their leaves and root systems, effectively purifying the surrounding atmosphere. Incorporating indoor plants into our living spaces can create a more natural and healthy environment.

12. Limiting Smoking Indoors:

The dangers of tobacco smoke have been extensively studied, and it is well-established that smoking indoors can have severe health consequences for both smokers and non-smokers. By designating outdoor areas for smoking and enforcing a smoke-free policy indoors, we can significantly reduce the presence of harmful chemicals in our breathing spaces.

13. Source Control:

One of the most effective ways to minimize exposure to airborne chemicals is through source control. By identifying and eliminating or reducing the use of products that emit harmful substances, we take a proactive approach to protect our health. This can involve selecting low or no VOC paints, opting for fragrance-free personal care products, or choosing natural alternatives to common household cleaners.

14. Regular Ventilation Maintenance:

To ensure that ventilation systems are functioning optimally, regular maintenance and cleaning are crucial. Dirty or clogged ventilation systems can hinder proper airflow and contribute to the build-up of airborne pollutants. By scheduling routine inspections and cleanings, we can promote the efficient and effective removal of pollutants from our indoor environments.

15. Education and Awareness:

Finally, empowering ourselves with knowledge about the potential risks associated with airborne chemicals is essential. By staying informed about the sources and health effects of these substances, we can make informed choices and advocate for change in our communities. Raising awareness among family, friends, and colleagues can also have a ripple effect, as collective action can lead to greater emphasis on healthier practices and policies.

In conclusion, by implementing these strategies to minimize airborne chemical exposure, we can create safer and healthier living spaces for ourselves and future generations. Awareness, coupled with proactive measures, is vital in unraveling the menace of airborne chemicals and ensuring a better quality of life. As we strive to protect our well-being, it is imperative to remember that each decision we make, no matter how small, can contribute to a healthier and more sustainable environment. Stay vigilant, and let us continue our journey towards a cleaner, safer world.

Chapter 5: Health Impacts of Airborne Chemicals

Diving into the detrimental effects on human health caused by exposure to airborne chemicals, this chapter explores both acute and chronic consequences. Airborne chemicals have become a growing concern in our modern world, as they infiltrate our environment and pose a serious threat to our well-being.

Air pollution, both indoors and outdoors, is a major contributor to the presence of airborne chemicals. Common sources include industrial emissions, vehicle exhaust, volatile organic compounds (VOCs) from cleaning products, and even everyday activities like cooking. As these pollutants fill the air we breathe, they can enter our bodies through inhalation, posing immediate and long-term risks to our health.

Acute effects of exposure to airborne chemicals can manifest soon after inhalation or contact. These effects vary depending on the specific chemical involved, its concentration, and the duration of exposure. Short-term symptoms may include respiratory problems, irritation of the eyes, nose, and throat, headaches, dizziness, and fatigue. Certain individuals may even experience allergic reactions or respiratory distress in severe cases.

Frequent exposure to airborne chemicals can lead to the development of chronic health conditions. Prolonged inhalation or absorption of harmful substances into the body over an extended period can have serious implications. Studies have linked chronic exposure to airborne chemicals with respiratory diseases such as asthma, chronic obstructive pulmonary disease (**COPD**), and lung cancer. Furthermore, certain chemicals have been associated with cardiovascular issues, neurodevelopmental disorders, and even reproductive abnormalities.

One group particularly vulnerable to the health impacts of airborne chemicals is children. Their developing bodies and organs are more susceptible to the toxic effects of these substances. Exposure during critical stages of growth and development can have long-lasting effects on their health and well-being. Moreover, their smaller lung size and higher respiratory rates make them more prone to inhalation of harmful substances present in the air.

The true extent of the health consequences caused by airborne chemicals is still being uncovered. Many studies have indicated potential links between exposure to certain chemicals and various health conditions, but further research is needed to fully understand the mechanisms at play. The complexity lies in the fact that humans are exposed to a combination of chemicals, making it challenging to isolate specific causes and effects.

Efforts to mitigate the health impacts of airborne chemicals involve both individual and collective actions. On an individual level, ensuring proper ventilation in homes and workplaces, using air purifiers, and minimizing exposure to known sources of indoor pollutants can contribute to reducing risk. Government regulations and policies play a crucial role in controlling emissions from industries, promoting cleaner technologies, and monitoring air quality to safeguard public health.

Understanding the health impacts of airborne chemicals is essential, as it allows us to take proactive measures to protect ourselves and future generations. Awareness and education are vital in empowering individuals to make informed decisions regarding their surroundings and lifestyle choices.

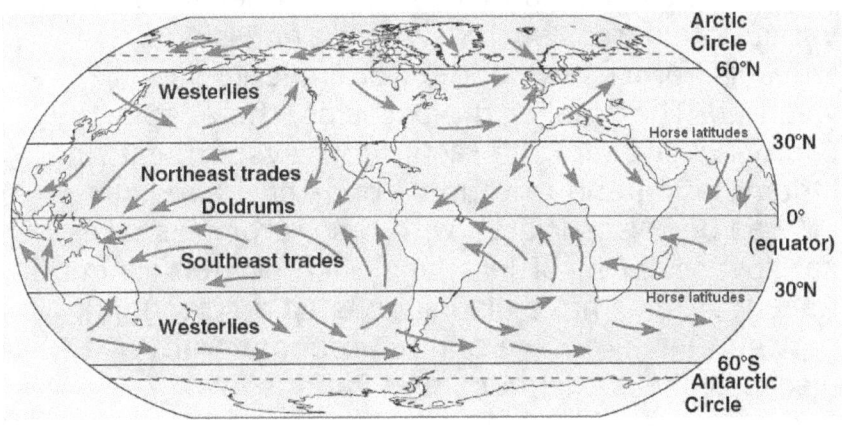

As we move forward in this chapter, we will delve deeper into specific airborne chemicals of concern and explore their effects on human health. By unraveling the complexities of these threats, we hope to shed light on the importance of addressing this widespread issue. The second half of this chapter will reveal key findings and strategies to mitigate the risks presented by airborne chemicals, ensuring a healthier and safer environment for all.

In exploring the effects of airborne chemicals on human health, it becomes evident that the impact is not limited to acute symptoms. Chronic exposure to these harmful substances can lead to a multitude of health conditions and long-term consequences. This second half of the chapter will further delve into specific airborne chemicals and their effects on human health. By unraveling the complexities of these threats, we aim to shed light on the importance of addressing this widespread issue to ensure a healthier and safer environment for all.

One of the most concerning airborne chemicals is particulate matter (PM). Particulate matter consists of a mixture of solid and liquid particles suspended in the air. These tiny particles can be released from a variety of sources such as vehicle emissions, industrial processes, and natural events like wildfires. The smaller the particles, the deeper they can penetrate into the respiratory system, causing inflammation and irritation. Long-term exposure to PM has been associated with respiratory diseases, cardiovascular disorders, and even premature death. Recent studies have also suggested a strong link between PM exposure and neurodegenerative diseases like Alzheimer's and Parkinson's.

Another group of airborne chemicals that poses a significant health risk is volatile organic compounds (**VOCs**). VOCs are emitted as gases from various household products, including cleaning supplies, paint, and building materials. Prolonged exposure to high concentrations of VOCs can have detrimental effects on both physical and neurological health. These compounds have been linked to respiratory conditions, allergies, and headaches. Additionally, some VOCs have been classified as carcinogens, with long-term exposure increasing the risk of developing cancer.

Furthermore, the presence of airborne heavy metals is a growing concern in many regions. Heavy metals such as lead, mercury, and arsenic are released from industrial activities, mining operations, and the burning of fossil fuels. Inhalation of these toxic metals can lead to severe health issues, including damage to the nervous system, kidney and lung damage, and developmental disorders in children. Even at low levels of exposure, heavy metals accumulate in the body over time, increasing the risk of chronic diseases.

Indoor air pollution, though often overlooked, can be equally—if not more—detrimental to human health. Many homes and workplaces harbor a variety of pollutants, including formaldehyde, radon, and mold. Formaldehyde is a VOC commonly found in building materials, furniture, and consumer products. Exposure to formaldehyde can cause respiratory irritation, allergic reactions, and, in some cases, contribute to the development of cancer. Radon, a radioactive gas, is released naturally from the ground and can seep into homes through cracks and gaps. Prolonged exposure to high levels of radon has been linked to an increased risk of lung cancer. Mold growth, often triggered by high humidity or water damage, releases spores that can exacerbate respiratory conditions and cause allergies.

Children are particularly vulnerable to the health impacts of airborne chemicals. Their developing bodies and organs are more susceptible to the toxic effects of these substances. Moreover, their smaller lung size and higher respiratory rates make them more prone to inhaling harmful pollutants present in the air. Studies have shown that early-life exposure to air pollutants can lead to impaired lung function, neurodevelopmental delays, and lifelong health issues. Action must be taken to protect the most vulnerable populations from the harmful effects of airborne chemicals.

To mitigate the risks presented by airborne chemicals, various strategies have been implemented at both individual and collective levels. On an individual level, ensuring proper ventilation in homes and workplaces, using air purifiers, and minimizing exposure to known sources of indoor pollutants can significantly reduce risk. It is crucial to be aware of potential sources of airborne chemicals, such as household products, and opt for safer alternatives whenever possible.

At the collective level, government regulations and policies play a vital role in controlling emissions from industries, promoting cleaner technologies, and monitoring air quality to safeguard public health. Cooperation between government bodies, industries, and communities is necessary to establish and enforce stringent standards for air quality. Additionally, public awareness campaigns and education initiatives are crucial in empowering individuals to make informed decisions and take proactive steps in creating a healthy environment.

In conclusion, the detrimental effects of airborne chemicals on human health are a pressing concern in today's world. Chronic exposure to these harmful substances can lead to a wide range of health conditions and long-term consequences. Understanding the specific chemicals involved, such as particulate matter, VOCs, and heavy metals, is essential for developing effective strategies to mitigate the risks. By addressing this issue collectively and implementing measures at both individual and collective levels, we can create a healthier and safer environment for ourselves and future generations.

Chapter 6: The Silent Spread: Routes of Dispersion

Examining the various ways airborne chemicals disperse, this chapter unravels how they can travel across vast distances and infiltrate unsuspecting environments. In today's world, where industrial activities and technological advancements have become an integral part of our lives, the presence and potential danger of airborne chemicals cannot be ignored. Understanding the routes of dispersion is crucial to grasping the full extent of this silent menace.

One of the primary routes through which airborne chemicals spread is the atmosphere. As we go about our daily lives, we release a wide range of substances into the air, whether through industrial processes, transportation emissions, or even simple household activities. Once released, these chemicals become part of our atmosphere and can travel great distances under the influence of wind patterns and atmospheric conditions.

Wind plays a significant role in determining the direction and speed at which airborne chemicals disperse. Constant air currents around the globe form part of a complex system known as atmospheric circulation. These air currents, such as trade winds and westerlies, create atmospheric highways that facilitate the movement of airborne chemicals beyond their source locations. Once caught in the flow, these chemicals can journey across countries, continents, and even oceans, defying any boundaries or borders.

The size and weight of airborne chemicals also influence their dispersion. Those that are smaller and lighter, such as volatile organic compounds (VOCs), have a higher chance of remaining suspended in the air for extended periods. This enables them to travel longer distances before eventually settling or undergoing chemical transformations. In contrast, larger or heavier particles, such as dust or pollen, tend to settle closer to their source, with limited dispersion capabilities.

Another route of dispersion lies in the interconnectedness of ecosystems. Nature, with its delicate balance of interconnected systems, offers opportunities for airborne chemicals to navigate vast distances. These chemicals hitchhike on natural processes, such as water cycles, animal migrations, and even the flight patterns of migratory birds.

Water cycles, for instance, can carry airborne chemicals through the process of evaporation, condensation, and precipitation. Chemicals that dissolve easily in water, like mercury or certain pesticides, can be transported across regions as they adhere to water droplets. This process not only affects aquatic environments but also influences the quality of air in remote areas far from original sources.

Animal migrations and their flight patterns allow chemicals to reach unexpected and remote habitats. Birds, for instance, cover thousands of miles during their migration routes, potentially introducing pollutants from one location to another. The silent spread of chemicals is amplified by the far-reaching movements of these migratory species, as they unknowingly become transporters of contamination.

Furthermore, human activities like deforestation and urbanization have disrupted ecosystems, altering natural dispersion patterns and intensifying chemical spreading. By destroying natural barriers and creating pathways for pollutants, we have inadvertently supercharged the silent spread of airborne chemicals.

Consider the scenario of a remote forest located miles away from an industrial area. Through the complex interplay of wind patterns, atmospheric circulation, and the interconnectedness of ecosystems, toxic chemicals released from industries can find their way to this serene forest. From the towering trees to the smallest organisms that dwell within, this once-pristine ecosystem becomes vulnerable to the silent intrusion of harmful substances.

The silent spread of airborne chemicals poses a grave threat to both human and environmental health. It is through understanding the various routes of dispersion that we can begin to comprehend the scope of this menace. By unravelling the complexities of atmospheric circulation, the influence of wind patterns, and the interconnectedness of ecosystems, we take the first step towards mitigating the risks associated with airborne chemical pollution.

In the second half of this chapter, we will delve deeper into the specific ways these airborne chemicals infiltrate unsuspecting environments, exploring their impacts and potential solutions. Join us as we shine a light on the hidden perils that lurk beneath the surface, awaiting their opportunity to strike. The journey into the heart of the silent spread continues, unravelling the intricate web of dispersion and infiltration. The quest for knowledge and protection against this invisible threat has only just begun.As we navigate further into the intricate web of dispersion and infiltration, we encounter the specific ways in which airborne chemicals infiltrate unsuspecting environments, unraveling their impacts and potential solutions. It is through a comprehensive understanding of these processes that we can work towards mitigating the risks associated with airborne chemical pollution.

One of the primary mechanisms through which airborne chemicals penetrate different environments is atmospheric deposition. This refers to the process by which particles and gases settle onto surfaces, including soil, water bodies, plants, and human-made structures. These chemicals can be directly deposited from the atmosphere or indirectly through precipitation, commonly known as wet deposition. Rain, snow, and fog can carry contaminants as they precipitate, effectively "washing" the chemicals from the air onto the ground.

The effects of atmospheric deposition are most significant in areas downwind of pollution sources. Industrial regions, densely populated cities, and areas close to major transportation routes often experience higher levels of deposition. As these chemicals settle, they become integrated into the environment, potentially posing risks to ecosystems, wildlife, and human health.

Once airborne chemicals have infiltrated the environment, their impacts can be diverse and far-reaching. For instance, vegetation and crops can suffer adverse effects from the deposition of pollutants onto their leaves and needles. The chemicals may inhibit photosynthesis, disrupt nutrient absorption, or impair the overall health and growth of plants. This has ramifications for agricultural productivity, food security, and biodiversity.

Similarly, aquatic ecosystems are vulnerable to the silent spread of airborne chemicals. Toxic substances can enter lakes, rivers, and oceans through atmospheric deposition or runoff from contaminated soils. Aquatic organisms, such as fish and other aquatic species, can be directly affected by these chemicals, leading to physiological disorders, reproductive issues, and even population decline. Moreover, the accumulation of pollutants in the food chain poses a threat to the health of humans who consume contaminated seafood.

Human health is also jeopardized by the infiltration of airborne chemicals into our environments. People can be exposed to these pollutants through inhalation, ingestion, or dermal contact with contaminated surfaces. The health effects range from respiratory issues, neurological disorders, and cardiovascular diseases to an increased risk of cancer and reproductive abnormalities. Vulnerable populations, such as children, the elderly, and individuals with pre-existing health conditions, are particularly susceptible.

To tackle the hazards associated with airborne chemical pollution, a multi-faceted approach is necessary. Firstly, stringent regulations on industrial emissions and transportation sectors can help reduce the release of harmful substances into the air. Encouraging the adoption of cleaner technologies, sustainable practices, and the use of alternative energy sources can contribute to minimizing the silent spread.

Public awareness and education also play a vital role in combating the menace of airborne chemicals. By disseminating information about the risks associated with pollution and the steps individuals can take to mitigate their own impact, we empower people to make conscious choices and promote collective responsibility.

Additionally, environmental monitoring and research are crucial to assessing the extent of contamination and understanding its ecological and human health implications. By continuously monitoring air quality, water bodies, and soil samples, scientists can detect emerging pollutants, identify pollution sources, and develop effective strategies for pollution prevention and control.

Furthermore, restoration and remediation efforts are imperative in areas already affected by the silent spread of pollutants. Restoring ecosystems, implementing sustainable land management practices, and employing innovative techniques, such as phytoremediation (using plants to remove contaminants), can help revitalize contaminated environments and minimize the long-term impacts on both nature and human well-being.

In conclusion, the silent spread of airborne chemicals infiltrates unsuspecting environments through various routes of dispersion and infiltration. Atmospheric deposition and the impacts of contamination on vegetation, aquatic ecosystems, and human health illustrate the gravity of this invisible threat. However, with informed action, public awareness, and collective responsibility, we can work towards mitigating the risks associated with airborne chemical pollution. Through stringent regulations, education, monitoring, and restoration efforts, we can pave the way for cleaner, healthier environments, ensuring a sustainable future for ourselves and the planet we call home.

The journey into the heart of the silent spread has shed light on the crucial elements that contribute to the menace of airborne chemicals. As we take steps to address this issue, let us remain vigilant, informed, and committed to protecting the air we breathe, the water we drink, and the Earth we share.

Chapter 7: Indoor Air Quality: A Hidden Concern

Shedding light on the often overlooked issue of indoor air pollution, this chapter reveals the potential dangers lurking within our own homes and workplaces.

In today's fast-paced world, where we spend the majority of our time indoors, it's easy to forget that the air we breathe indoors can be just as harmful, if not more, than the air outside. We often associate pollution with belching factory smokestacks or congested city streets, but little do we realize that the very places we consider safe havens may harbor unseen threats.

Indoor air pollution refers to the presence of harmful substances indoors, which can significantly affect our health, comfort, and overall well-being. While it may be tempting to believe that our well-sealed homes and offices protect us from external pollutants, the truth is quite the contrary. In fact, indoor air can be up to five times more polluted than outdoor air, according to the Environmental Protection Agency (EPA). This revelation calls for immediate attention and action.

Numerous sources contribute to the deterioration of indoor air quality, and while some may be obvious, many are not. One of the leading causes is inadequate ventilation, resulting in stagnant air that allows pollutants to build up over time. Paints, varnishes, and cleaning products release volatile organic compounds (VOCs), which can lead to respiratory problems, headaches, and even cancer.

As we delve deeper into the realm of indoor air quality, we discover that even seemingly innocent activities can be culprits. Consider the cozy ambiance of a wood-burning fireplace or stove – a charm often enjoyed on chilly evenings. However, the soot and smoke emitted from burning wood release harmful particles, including carbon monoxide and fine particulate matter. These pollutants can cause respiratory issues, heart problems, and, in severe cases, even death.

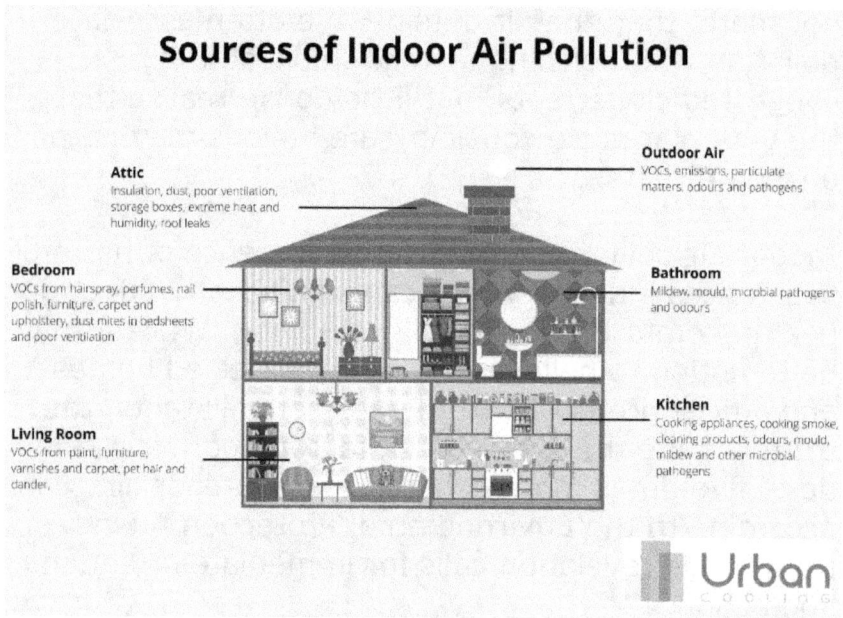

But it's not only indoor activities that contribute to the contamination of our breathing space. Outdoor pollutants, such as vehicle exhaust, pesticides, and industrial toxins, can easily infiltrate our homes through open windows and doors or via improperly sealed exteriors. The constant influx of allergens and pollutants from the outside can exacerbate existing health conditions or even trigger new ones.

To make matters worse, we often underestimate the role of everyday items in compromising indoor air quality. Furniture, carpets, and building materials emit harmful chemicals known as semi-volatile organic compounds (SVOCs). These substances not only affect our immediate health but have also been linked to long-term health problems, including hormone disruption and neurodevelopmental disorders.

In recent years, the prevalence of indoor air pollution has gained attention, prompting researchers, health professionals, and regulatory bodies to address this hidden menace. Organizations like the EPA have implemented guidelines to improve indoor air quality by focusing on ventilation, reducing exposure to hazardous substances, and promoting healthier building materials.

Awareness regarding the importance of maintaining good indoor air quality is growing, and individuals are increasingly taking measures to protect their living and working environments. This includes regular cleaning and maintenance, using air purifiers, opting for natural and eco-friendly products, and ensuring proper ventilation to allow fresh outdoor air to circulate.

While progress is being made, more research and action are required to fully understand the risks associated with indoor air pollution. It's not enough to simply assume that our indoor spaces are safe.

We must remain vigilant and proactive in safeguarding ourselves and our loved ones from the dangers that may silently permeate the air we breathe.

As we proceed with this chapter, we will explore the specific pollutants present in indoor environments, the associated health risks, and the steps individuals and society can take to mitigate this silent spread of airborne chemicals. Brace yourself, for the second half of this chapter will unravel a web of surprises and unveil solutions that can help ensure the air we breathe within our sanctuaries is truly safe. The second half of this chapter will dive deeper into the specific pollutants that contaminate indoor environments, the associated health risks, and the crucial steps individuals and society can take to mitigate the silent spread of airborne chemicals.

One of the major contributors to indoor air pollution is the combustion of fossil fuels, primarily through the use of gas stoves and heaters. These appliances release nitrogen dioxide (NO_2), a highly toxic gas that can lead to respiratory problems, particularly in vulnerable individuals such as children and the elderly. Long-term exposure to NO_2 has been linked to an increased risk of respiratory infections, asthma, and even lung cancer. Proper ventilation and regular maintenance of gas appliances are essential to minimize the release of this harmful gas into the indoor air.

Additionally, the presence of mold and dampness in buildings is a pervasive problem that can significantly impair indoor air quality. Mold spores that are released into the air can trigger allergic reactions, respiratory issues, and even contribute to the development of asthma. To tackle this issue, it is crucial to promptly address any leaks or areas of moisture accumulation in homes and offices. Adequate ventilation and maintaining indoor humidity levels below 50 percent can help prevent the growth of mold and improve overall air quality.

Another overlooked source of indoor air pollution is the use of certain cleaning products. Many conventional cleaning agents contain volatile organic compounds (VOCs) that are released into the air during and after use. These chemicals can irritate the respiratory system, trigger asthma attacks, and even have long-term health effects. Opting for eco-friendly cleaning products or homemade alternatives can significantly reduce exposure to harmful VOCs and promote better indoor air quality.

The role of building materials in contributing to indoor air pollution cannot be ignored. Many common building materials, such as paints, adhesives, and furniture, release toxic chemicals like formaldehyde and benzene into the air. These substances can cause irritation of the eyes, throat, and respiratory system, and long-term exposure may increase the risk of cancer. When renovating or furnishing indoor spaces, it is essential to choose low-VOC or VOC-free products. Proper ventilation during and after construction or renovation can also facilitate the removal of harmful emissions.

In recent years, electronic devices such as printers, copiers, and computers have become ubiquitous in both residential and office settings. However, the materials used in these devices, as well as the ink and toner cartridges, can emit hazardous substances like ozone, particulate matter, and volatile chemicals. Prolonged exposure to these pollutants can lead to respiratory issues, allergies, and even cardiovascular problems. Regular cleaning and maintenance of electronic devices, along with proper ventilation, can help minimize the release of these pollutants and ensure healthier indoor air.

One of the most effective ways to improve indoor air quality is by increasing ventilation rates. Opening windows and doors, using exhaust fans, and ensuring a constant flow of fresh air can dilute and remove indoor pollutants. Additionally, the installation of mechanical ventilation systems, such as whole-house ventilation or air purifiers, can efficiently filter out pollutants and maintain a healthy air exchange rate.

Individuals can also take proactive steps to protect themselves from indoor air pollution. Regularly cleaning and dusting indoor areas, using vacuum cleaners with HEPA filters, and keeping indoor humidity levels low can significantly reduce the presence of allergens and particulate matter in the air. Avoiding smoking indoors and discouraging the use of harmful chemicals and strong fragrances also contribute to better indoor air quality.

On a broader scale, regulatory bodies and organizations are taking steps to address the menace of indoor air pollution. The Environmental Protection Agency (EPA) has established guidelines and regulations for indoor air quality in various settings. These guidelines include recommendations for improving ventilation, reducing exposure to hazardous substances, and promoting the use of healthier building materials.

In conclusion, the second half of this chapter has shed light on the specific pollutants that contaminate indoor environments, the associated health risks, and the necessary steps individuals and society can take to combat indoor air pollution. With a deeper understanding of the sources and effects of airborne chemicals, it is crucial to remain vigilant and proactive in safeguarding our indoor spaces. By adopting healthier practices, advocating for better regulations, and utilizing available technologies, we can ensure that the air we breathe within our homes and workplaces is truly safe and conducive to our well-being.

Chapter 8: Airborne Chemical Regulation: A Global Perspective

Delving into the existing regulatory frameworks and international efforts to control airborne chemicals, this chapter presents an overview of global initiatives and their effectiveness. The silent spread of airborne chemicals is a growing concern, with potentially devastating consequences for human health and the environment. As our world becomes increasingly interconnected, understanding the global perspective on regulating these substances is paramount.

Airborne chemical pollution knows no borders. The pollutants released into the atmosphere can travel across vast distances, affecting regions far from their original sources. Recognizing the urgent need for comprehensive regulation, numerous international efforts have been put forth to address this pervasive issue.

At the international level, the United Nations plays a vital role in coordinating efforts to regulate airborne chemicals. The Stockholm Convention on Persistent Organic Pollutants (POPs) stands as one of the most important global agreements in this domain. Adopted in 2001, the convention aims to eliminate or restrict the production, use, and release of POPs, which possess toxic properties, persist in the environment, and bioaccumulate in living organisms, including humans.

The inclusion of certain chemicals in the Stockholm Convention's list of POPs, such as polychlorinated biphenyls (PCBs) and dioxins, highlights the grave risks associated with exposure to these substances. By providing a legally binding framework for member countries, the convention establishes a strong foundation for coordinated action against the silent spread of airborne pollutants.

Several other international organizations and initiatives supplement the efforts initiated by the Stockholm Convention. The World Health Organization (WHO) actively collaborates with governments and non-governmental organizations to assess the health impacts of airborne chemicals and develop guidelines for exposure limits. Through its International Agency for Research on Cancer (IARC), the WHO evaluates the carcinogenicity of various substances, shedding light on the potential risks associated with their presence in the air we breathe.

The United Nations Environment Programme (UNEP) also plays a pivotal role in the global regulation of airborne chemicals. UNEP's focus extends beyond persistent pollutants, encompassing a wider array of chemical substances that pose risks to human health and the environment. Through regional initiatives, such as the Minamata Convention on Mercury and the Basel Convention on the Control of Transboundary Movements of Hazardous Wastes and Their Disposal, UNEP strives to develop comprehensive strategies for minimizing the release and impact of airborne chemicals.

Beyond these international agreements, regional efforts have emerged as centers of regulatory action. The European Union (EU), for instance, has solidified its commitment to regulating airborne chemicals by strict legislation, such as the Registration, Evaluation, Authorization, and Restriction of Chemicals (REACH) regulation. REACH establishes a robust system for the registration, evaluation, and authorization of chemical substances within the EU market, enhancing transparency, accountability, and public safety.

Additionally, the North American region spearheads efforts through the North American Agreement on Environmental Cooperation (NAAEC). This trilateral agreement between Canada, Mexico, and the United States focuses on addressing environmental concerns, including the regulation of airborne chemicals. By fostering cooperation and sharing best practices, NAAEC exemplifies the significance of collaboration in combating the menace of airborne pollutants.

While these global, regional, and national initiatives mark significant progress, challenges persist. The effectiveness of regulatory frameworks relies heavily on compliance and enforcement mechanisms, as well as the sharing of scientific data and expertise. Harmonizing regulations between countries with varying capacities and priorities remains a complex task. Furthermore, the rapid pace of scientific advancements necessitates the continuous updating and improvement of regulatory frameworks to keep pace with emerging airborne chemical risks.

As we delve deeper into the second half of this chapter, we will explore the specific mechanisms and challenges associated with global and regional regulatory frameworks. By analyzing case studies and evaluating the effectiveness of existing initiatives, we aim to uncover potential areas for improvement and outline a path forward in the fight against the silent spread of airborne chemicals.

And with that, the first half of this chapter concludes, leaving us on the cusp of a deeper understanding of global regulatory efforts. The journey continues as we unveil the intricacies and potential solutions to the challenges faced in controlling these invisible yet pervasive threats. Stay tuned for the second half of Chapter 8, where we will embark on a comprehensive exploration of the international landscape of airborne chemical regulation. As we delve deeper into the second half of this chapter, we continue our exploration of the specific mechanisms and challenges associated with global and regional regulatory frameworks. Through the analysis of case studies and the evaluation of existing initiatives, we aim to uncover potential areas for improvement and outline a path forward in the fight against the silent spread of airborne chemicals.

One of the key challenges in regulating airborne chemicals lies in the diverse capacities and priorities of different countries. Harmonizing regulations across borders is a complex task that requires international collaboration and agreement. However, the effectiveness of regulatory frameworks depends heavily on compliance and enforcement mechanisms. Without strong enforcement, regulations may remain merely on paper, failing to effectively protect human health and the environment.

To address this challenge, international organizations such as the United Nations and its subsidiary agencies play a crucial role in facilitating cooperation and promoting compliance. The United Nations

Environment Programme (**UNEP**), for instance, supports countries in implementing and enforcing international agreements on airborne chemicals. Through capacity-building programs and technical assistance, UNEP assists nations in strengthening their regulatory frameworks and monitoring systems.

Moreover, enhancing the sharing of scientific data and expertise is vital in improving the effectiveness of airborne chemical regulation. The dissemination of knowledge and research findings allows countries to make informed decisions and develop evidence-based policies. Efforts to establish global databases on chemical information, such as the Global Monitoring Plan under the Stockholm Convention, contribute to this knowledge-sharing endeavor.

Furthermore, the rapid pace of scientific advancements necessitates the continuous updating and improvement of regulatory frameworks. As scientific understanding evolves, new airborne chemical risks may emerge, requiring prompt action. Collaborative platforms and mechanisms, such as the Intergovernmental Forum on Chemical Safety and the International Pollutants Elimination Network (IPEN), facilitate the exchange of scientific knowledge and best practices among policymakers, regulators, and researchers, supporting the ongoing refinement of regulatory approaches.

Case studies provide valuable insights into the effectiveness of existing initiatives and highlight areas for improvement. The Minamata Convention on Mercury, for example, exemplifies international efforts to address the specific risks associated with this pollutant. By regulating the production, use, and emission of mercury, the convention aims to protect human health and the environment from its toxic effects. However, challenges remain in implementing the convention's provisions, particularly in developing countries with limited resources and capacity.

Similarly, the Basel Convention on the Control of Transboundary Movements of Hazardous Wastes and Their Disposal addresses the global management of hazardous wastes, including airborne chemicals. By establishing guidelines for the transboundary movement of such wastes and promoting their environmentally sound disposal, the convention aims to prevent harm to human health and the environment. However, issues such as illegal trafficking of hazardous wastes and the lack of uniformity in national regulations pose challenges to its effective implementation.

Regional initiatives also contribute significantly to the regulation of airborne chemicals. The European Union's stringent legislation, such as the REACH regulation, sets a benchmark for transparency, accountability, and public safety. By requiring the registration and evaluation of chemical substances, REACH enhances the understanding of their potential risks and enables appropriate risk management measures. The success of such regional approaches highlights the importance of collaboration and shared responsibility in combating the menace of airborne pollutants.

The North American region, through the North American Agreement on Environmental Cooperation (**NAAEC**), demonstrates the benefits of trilateral cooperation in addressing environmental concerns, including airborne chemical regulation. By fostering dialogue, sharing best practices, and harmonizing approaches, NAAEC promotes effective regulation while respecting each country's sovereignty and priorities.

Despite the progress made by global and regional initiatives, ongoing challenges persist. Stronger enforcement mechanisms, increased financial support, and robust monitoring systems are necessary to ensure compliance and measure the impact of regulatory actions. Additionally, further research and development of alternative, safer chemicals and technologies are essential to reducing reliance on harmful substances.

In conclusion, the regulation of airborne chemicals requires a global perspective and collaborative approach. International efforts, exemplified by the Stockholm Convention and the work of organizations like UNEP and WHO, provide a foundation for action. Regional initiatives, such as those undertaken by the European Union and North America, offer valuable insights into effective regulation. However, challenges in compliance, enforcement, and scientific advancements persist. By understanding the mechanisms and limitations of existing frameworks, we can identify areas for improvement and pave the way for a safer and healthier future. Stay tuned for the next chapter, where we will delve into the innovative strategies and groundbreaking research shaping the future of airborne chemical regulation.

Chapter 9: Case Studies: Notable Airborne Chemical Disasters

Investigating major incidents of airborne chemical disasters, this chapter explores the profound and far-reaching impacts they have had on communities and ecosystems. From devastating industrial accidents to chemical warfare incidents, the release of airborne chemicals has brought unimaginable chaos and devastation throughout history. By examining some notable case studies, we can truly comprehend the magnitude of these disasters and the urgent need for effective preventive measures.

One of the most infamous airborne chemical disasters occurred in December 1984 in **Bhopal, India**. The Union Carbide pesticide plant released approximately 40 metric tons of highly toxic methyl isocyanate (MIC) gas into the surrounding environment. This disaster immediately claimed thousands of lives and left countless others struggling with severe health issues in the aftermath. The incident not only showcased the devastating short-term effects of airborne chemical disasters but also highlighted the long-term impacts on the affected individuals and the environment.

Further illustrating the catastrophic consequences of airborne chemical disasters is the **Chernobyl nuclear accident** that took place in April 1986. While primarily known for its radioactive fallout, this disaster also resulted in the release of various airborne toxic chemicals, including dioxins and polychlorinated biphenyls (PCBs). These chemicals contaminated the air, soil, and water, impacting the health of nearby communities and the surrounding ecosystem. The Chernobyl disaster serves as a grim reminder of the interconnectedness between different types of hazards and the multiple risks posed by airborne chemical disasters.

Moving forward in time, the Deepwater Horizon oil spill in April 2010 revealed the devastating impact that airborne chemicals can have on marine environments. The release of an estimated 4.9 million barrels of crude oil into the Gulf of Mexico not only affected marine life directly but also led to the dispersal of volatile organic compounds (VOCs) into the air. These VOCs not only posed immediate health risks to cleanup workers but also had long-lasting effects on coastal communities and ecosystems, including the disruption of delicate ecological balances and the contamination of surrounding air.

While these case studies demonstrate the severity of airborne chemical disasters, it is important to note that such incidents are not limited solely to industrial accidents or technological failures. Chemical warfare has also left a harrowing imprint on history, with significant ramifications on both human lives and the environment.

One haunting example is the use of chemical weapons during World War I. From the chlorine gas attacks in 1915 to the even deadlier employment of mustard gas and phosgene, these chemical agents ravaged the battlefields, causing immense suffering and death. The airborne nature of these chemicals allowed them to spread indiscriminately, impacting not only the soldiers but also civilians residing near the frontlines. The consequences of these chemical warfare incidents were both immediate and long-lasting, leaving a lasting scar on the collective memory of humanity.

As we delve into the second half of this chapter, we will explore additional case studies that further illustrate the grave repercussions of airborne chemical disasters. From more recent incidents like the Seveso disaster in Italy to the ongoing challenges presented by the destruction of chemical weapons, each story serves as a sobering reminder of the importance of proactive measures and international cooperation in preventing and mitigating such catastrophes.

In the face of these disasters, it becomes evident that the effects of airborne chemicals extend beyond physical harm. The psychological impact on affected communities, the economic ramifications, and the long-term environmental consequences all require careful consideration and analysis. By understanding the full scope of these disasters, we can work towards developing effective strategies to prevent future occurrences and ensure the safety and well-being of our communities.

One shocking example of an airborne chemical disaster is the tragedy that occurred in **Minamata, Japan**, during the mid-1950s. The Chisso Corporation, a chemical company, released massive amounts of mercury compounds into the surrounding waters of Minamata Bay, contaminating the seafood consumed by the local population. The toxic mercury in the seafood led to a severe neurological syndrome known as Minamata disease, causing neurological impairments, paralysis, and even death.

The impacts of Minamata disease were devastating, both to the individuals affected and the ecosystem. Thousands of people suffered from the debilitating effects of mercury poisoning, and many lost their lives. The fishing industry in Minamata also collapsed, leaving local communities economically and socially devastated. The incident exposed the tragic consequences of environmental contamination and underscored the need for stringent regulations to prevent similar occurrences.

Continuing with a more recent case study, we turn to the West Virginia water crisis in January 2014. Residents of Charleston and surrounding areas were exposed to a chemical known as MCHM (4-methylcyclohexane methanol) when it leaked from a storage tank at a chemical facility along the Elk River. The chemical entered the region's water supply, affecting the drinking water of approximately 300,000 residents.

The incident sparked widespread panic and uncertainty as the full extent of the chemical's effects on human health remained unclear. The release of MCHM not only posed immediate health risks but also led to long-term concerns about potential chronic health effects. Moreover, the crisis exposed the vulnerabilities in chemical storage and safety regulations, calling attention to the need for stricter oversight and emergency response protocols.

Another case study that highlights the insidious nature of airborne chemical disasters is the Love Canal disaster, which unfolded in Niagara Falls, New York, in the late 1970s. In the 1940s and 1950s, Hooker Chemical Company used an abandoned canal as a disposal site for toxic waste, including chemicals such as benzene, dioxins, and polychlorinated biphenyls (PCBs). Over the years, the buried waste leaked into the soil and groundwater, contaminating the surrounding neighborhoods.

Residents of Love Canal began experiencing alarming health issues, including birth defects, developmental delays, and elevated rates of cancer. Children playing in the area developed rashes and suffered from mysterious illnesses. As the community raised concerns, it became clear that the chemicals buried beneath their feet were the cause.

The Love Canal disaster became a pivotal moment in environmental history, leading to the creation of the Superfund program to clean up hazardous waste sites. It also raised awareness about the potential hazards associated with improper waste disposal and the need for responsible industrial practices.

Lastly, we examine the devastating consequences of the Union Carbide chemical plant explosion in Texas City, Texas, in March 2005. A series of explosions and fires ripped through the facility, resulting in the release of large amounts of toxic chemicals, including highly carcinogenic substances like benzene. The incident not only tragically claimed the lives of 15 workers but also caused injuries to hundreds of others.

The Union Carbide explosion exposed the vulnerability of industrial sites and the potential for catastrophic events if proper safety measures are not in place. It revealed the importance of comprehensive risk assessment, emergency response plans, and rigorous safety protocols within the industrial sector to avoid such catastrophic events.

Meticulously studying these case studies of notable airborne chemical disasters expands our understanding of their far-reaching consequences. It emphasizes the urgent need for stringent regulations, effective preventive measures, and proactive international cooperation to mitigate, if not prevent, future occurrences. By respecting humanity's shared responsibility to protect the environment and prioritize the safety and well-being of the communities, we can strive towards a future where the silent spread of airborne chemicals is reduced, averting the immense suffering and devastation they inflict.

Chapter 10: Detecting and Monitoring Airborne Chemicals

Unveiling the sophisticated techniques and technologies used for detecting and monitoring airborne chemicals, this chapter reveals the tools available for identifying potential risks. Air pollution is a growing concern in today's world, and understanding how to effectively detect and monitor airborne chemicals is essential for safeguarding human health and the environment.

Airborne chemicals can originate from various sources, including industrial emissions, vehicle exhaust, agricultural activities, and household products. These chemicals can have detrimental effects on human health, leading to respiratory diseases, allergic reactions, and even long-term complications. Furthermore, they can also harm ecosystems, contaminating water bodies and affecting wildlife.

To effectively tackle this pressing issue, scientists and researchers have developed a range of sophisticated techniques and technologies for detecting and monitoring airborne chemicals. These tools are designed to help identify and assess the presence of hazardous substances in the air we breathe, enabling timely interventions and preventive measures.

One of the most common techniques used for detecting and monitoring airborne chemicals is air sampling. This involves collecting air samples from specific locations and analyzing them in specialized laboratories. Air samples can be collected using various methods, such as passive samplers, active samplers, and real-time monitoring devices. Passive samplers, like diffusion tubes, collect air pollutants over a specific period, providing a cumulative measurement. Active samplers, on the other hand, use a pump to actively pull air through a filter or absorbent media, allowing for the collection of a larger volume of air and more accurate measurements. Real-time monitoring devices provide instantaneous data and are often used for continuous monitoring in high-risk areas.

Once air samples are collected, various analytical techniques are employed to identify and quantify the presence of airborne chemicals. Gas chromatography-mass spectrometry (GC-MS) is a widely used technique that separates and analyzes chemical mixtures, providing information about the composition and concentration of different chemicals. Another commonly utilized technique is high-performance liquid chromatography (HPLC), which is particularly effective for analyzing organic compounds. Spectroscopic techniques, such as infrared spectroscopy and ultraviolet-visible spectroscopy, are also valuable tools for identifying specific chemical compounds based on their unique spectral signatures.

In recent years, advancements in technology have led to the development of innovative monitoring approaches. Remote sensing is one such technique that uses satellites and aerial platforms to detect and monitor airborne chemicals over large areas. Through the analysis of spectral data, remote sensing allows for the identification of pollutant sources, mapping of contamination patterns, and monitoring of changes in air quality over time. This technology provides valuable insights for environmental agencies and policymakers in implementing effective emission reduction strategies.

Additionally, sensor-based technologies have revolutionized the field of air quality monitoring. Small, portable sensors equipped with various detection mechanisms can be used to measure a wide range of airborne chemicals in real-time. These sensors provide high-resolution data and can be deployed in urban areas, industrial sites, or even worn by individuals for personal exposure monitoring. Advancements in miniaturization and wireless communication have made sensor networks a powerful tool for obtaining real-time air quality data with high spatial and temporal resolution.

The utilization of artificial intelligence (AI) and machine learning algorithms is also becoming increasingly prevalent in the field of detecting and monitoring airborne chemicals. AI-powered systems can analyze vast amounts of data collected from sensors, satellites, and other monitoring devices to identify patterns, predict pollutant levels, and model the dispersion of chemicals in the atmosphere. These advanced computational techniques enhance the accuracy and efficiency of air quality monitoring, enabling more informed decision-making.

As we delve deeper into the second half of this chapter, we will explore emerging technologies and their applications in detecting and monitoring airborne chemicals. The continuous advancements in this field hold promise for improving our understanding of air pollution and its impacts, leading us towards a cleaner and healthier future. Stay tuned for the next part, where we will unravel the exciting developments and potential solutions that lie ahead.

Advanced technologies have revolutionized the field of detecting and monitoring airborne chemicals, offering new avenues for mitigating the risks associated with air pollution. In this second half of the chapter, we will delve deeper into emerging technologies and their applications, providing valuable insights into the exciting developments that lie ahead.

One of the emerging technologies in the field of air quality monitoring is remote sensing. Satellites and aerial platforms equipped with sophisticated sensors have enabled the detection and monitoring of airborne chemicals over large areas. By analyzing spectral data, remote sensing allows for the identification of pollutant sources, mapping of contamination patterns, and monitoring of changes in air quality over time. This technology provides invaluable information for environmental agencies and policymakers in implementing effective emission reduction strategies.

Furthermore, sensor-based technologies have played a pivotal role in revolutionizing air quality monitoring. Small, portable sensors equipped with various detection mechanisms can now measure a wide range of airborne chemicals in real-time. These sensors provide high-resolution data and can be deployed in urban areas, industrial sites, or even worn by individuals for personal exposure monitoring. Advancements in miniaturization and wireless communication have made sensor networks a powerful tool for obtaining real-time air quality data with high spatial and temporal resolution. This wealth of data empowers decision-makers to develop targeted interventions to mitigate air pollution risks.

The integration of artificial intelligence (AI) and machine learning algorithms has greatly enhanced the accuracy and efficiency of detecting and monitoring airborne chemicals. AI-powered systems can analyze vast amounts of data collected from sensors, satellites, and other monitoring devices to identify patterns, predict pollutant levels, and model the dispersion of chemicals in the atmosphere. These advanced computational techniques enable more informed decision-making, allowing for the prompt implementation of corrective measures when necessary.

A promising application of AI in air quality monitoring is the development of predictive models for early warning systems. By analyzing historical data and continuously updating with real-time information, these models can identify and predict episodes of high air pollution levels. This timely information can alert authorities and communities to take preventive measures, such as implementing temporary emission controls or issuing health advisories to vulnerable populations.

Another area of ongoing research and development is the use of nanotechnology in air quality monitoring. Nanomaterial-based sensors offer improved sensitivity, selectivity, and stability, making them ideal for detecting trace amounts of airborne chemicals. These sensors can provide real-time data by capturing and analyzing air samples at a molecular level, enhancing our ability to identify and quantify hazardous substances accurately.

In recent years, the concept of smart cities has gained momentum, aiming to integrate urban infrastructure with the latest technological advancements for efficient resource management and improved quality of life. Air quality monitoring is a crucial component of smart city initiatives, utilizing real-time data from sensor networks, satellite imagery, and other sources to enable data-driven decision-making. The integration of air quality data into broader urban planning strategies allows for the implementation of targeted measures to improve air quality and protect public health.

As we move forward, innovative monitoring approaches continue to emerge. One such approach is the utilization of unmanned aerial vehicles (UAVs), commonly known as drones, for air quality monitoring. Equipped with specialized sensors, UAVs can provide detailed information about air pollution levels in areas that are difficult to access or monitor using conventional methods. The high mobility and flexibility of drones make them an ideal tool for identifying localized sources of pollution, monitoring industrial emissions, or assessing air quality during emergencies or natural disasters.

Additionally, citizen science initiatives have become increasingly popular, engaging the public in the process of collecting air quality data. Through the use of low-cost sensors and mobile applications, individuals can contribute to monitoring efforts by measuring air pollution levels in their surroundings. This participatory approach not only increases the spatial coverage of monitoring networks but also raises awareness and empowers communities to take action.

In conclusion, the second half of this chapter has explored the emerging technologies and applications in detecting and monitoring airborne chemicals. Remote sensing, sensor-based technologies, AI and machine learning algorithms, nanotechnology, and the integration of air quality monitoring into broader urban planning strategies are some of the exciting developments shaping the future of this field. As we continue to advance our understanding of air pollution and its impacts, harnessing these innovations holds great promise for creating a cleaner and healthier future. By leveraging the power of technology, we can identify potential risks, implement effective interventions, and protect human health and the environment.

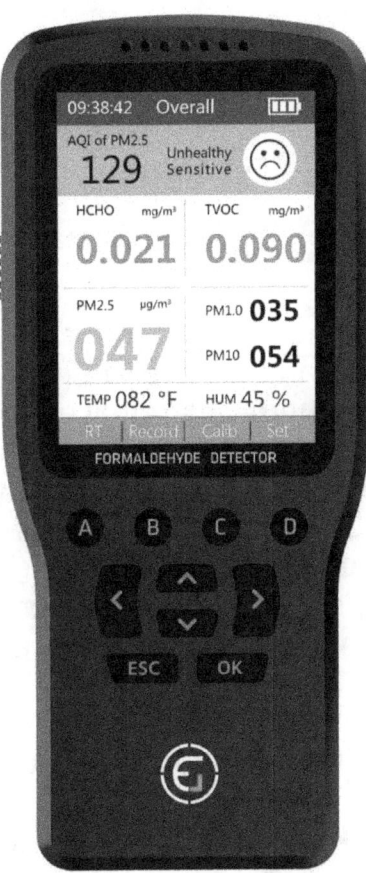

Chapter 11: Mitigating Airborne Chemical Exposure

Airborne chemicals can pose a significant threat to our health and the well-being of our communities. From the chemicals released in industrial processes to everyday pollutants found in our homes, these invisible substances can silently infiltrate our bodies, leading to various health issues and long-term consequences. However, with knowledge and practical strategies, we can empower ourselves and take proactive steps to reduce or eliminate airborne chemical exposure. In this chapter, we will explore effective methods to protect ourselves and our communities from this menacing threat.

1. Identify and Eliminate Potential Sources of Airborne Chemicals

The first step in mitigating airborne chemical exposure is identifying and eliminating the sources of these hazardous substances. Regularly inspect your home, workplace, and surroundings for potential sources such as cleaning products, pesticides, and construction materials. Opt for non-toxic alternatives whenever possible, and ensure proper ventilation to dilute the concentration of airborne chemicals. By taking a proactive stance, we can significantly reduce our exposure to harmful substances.

2. Enhance Indoor Air Quality

Given that we spend most of our time indoors, it is crucial to ensure clean and healthy indoor air quality. Investing in air purifiers with effective filtration systems can help remove harmful particles and airborne chemicals. Additionally, regularly cleaning and dusting surfaces in your home can minimize the buildup of pollutants. Keeping indoor humidity levels in check and avoiding excessive use of scented products can also contribute to improving indoor air quality and reducing chemical exposure.

3. Educate Yourself and Raise Awareness

To truly protect ourselves and our communities from airborne chemical exposure, knowledge is key. Stay informed about potential risks and educate yourself about different chemicals commonly found in products. Be aware of potential health effects they can cause and take precautions accordingly. By sharing this knowledge, we can collectively create a culture of awareness and advocate for safer alternatives in our communities.

4. Support Regulations and Policies

While individual efforts can make a difference, the magnitude of the airborne chemical problem calls for broader societal action. Support regulations and policies aimed at reducing the release of harmful chemicals into the air. Get involved in local initiatives, organizations, and advocacy groups that prioritize environmental health. By standing together, we can influence change at both local and global levels.

5. Engage in Sustainable Practices

Adopting sustainable practices in our daily lives not only benefits the environment but also helps mitigate airborne chemical exposure. Opting for organic and locally sourced products reduces the reliance on harmful chemicals in agriculture and food production. Embracing energy-efficient appliances and transportation helps reduce air pollution. By making conscious choices that align with sustainability, we contribute to a healthier future for ourselves and our communities.

6. Collaborate and Communicate

Addressing airborne chemical exposure requires collective action. Collaborate with your neighbors, friends, and community members to raise awareness and find solutions together. Share information about best practices and encourage others to join the cause. Utilize social media platforms or community forums to promote discussions on airborne chemical exposure and systematic changes needed to combat this issue. By fostering open communication, we can create a powerful network dedicated to protecting our health and environment.

By implementing these practical strategies, we can take significant strides in reducing or eliminating airborne chemical exposure. The power to safeguard ourselves and our communities from this silent spread lies in our hands. As we venture into the second half of this chapter, we will delve deeper into more advanced techniques and innovative approaches to fortify our defense against airborne chemical threats. Stay tuned for the next part, where we unveil groundbreaking solutions and further expand our knowledge on this pressing matter.

In the quest to mitigate airborne chemical exposure, we have explored practical strategies that empower us to protect ourselves and our communities. By identifying and eliminating potential sources of airborne chemicals, enhancing indoor air quality, educating ourselves and raising awareness, supporting regulations and policies, engaging in sustainable practices, and collaborating and communicating with others, we can make significant progress in reducing the silent spread of these harmful substances.

However, to fortify our defense against airborne chemical threats, it becomes imperative to delve deeper into more advanced techniques and innovative approaches. In this second half of the chapter, we will explore some groundbreaking solutions that can further expand our knowledge on this pressing matter.

7. Passive Air Filtration Systems

While investing in air purifiers can effectively remove airborne chemicals, another approach gaining recognition is passive air filtration systems. These innovative systems leverage natural processes like plants' ability to absorb pollutants through their leaves and roots. By incorporating indoor plants strategically into our living and working spaces, we can enhance air quality and reduce chemical exposure. Additionally, passive air filtration systems offer aesthetic benefits and contribute to creating a greener and healthier environment.

8. Personal Protective Equipment

In certain circumstances, personal protective equipment (PPE) can play a vital role in reducing airborne chemical exposure. While PPE is often associated with industrial or healthcare settings, individuals can also consider using it when engaging in activities that involve high-risk exposure to chemicals. Depending on the nature of the task, suitable PPE such as respirators or chemical-resistant gloves can provide an extra layer of protection. It is crucial to choose the right type of equipment based on the specific chemicals involved and to follow proper usage guidelines to maximize effectiveness.

9. Air Pollution Monitoring Devices

Having access to real-time information about air quality is beneficial for individuals and communities concerned about airborne chemical exposure. Air pollution monitoring devices are becoming increasingly accessible and affordable, allowing us to measure the levels of pollutants in our surroundings. By utilizing these devices, we can make informed decisions about outdoor activities, choose safer routes, and advocate for necessary actions to reduce pollution levels in our neighborhoods.

10. Hazard Communication and Labeling

Clear and accurate labeling of chemical products is crucial in ensuring individuals can make informed choices and minimize exposure. Governments and regulatory bodies play a significant role in mandating proper hazard communication and labeling standards. By supporting these initiatives and advocating for transparent labeling practices, we empower consumers to identify potentially harmful products and opt for safer alternatives.

11. Green Building Design

From workplace environments to residential structures, incorporating green building design principles can significantly reduce airborne chemical exposure. Green buildings prioritize sustainability and health by utilizing eco-friendly construction materials, enhancing natural ventilation, and optimizing energy efficiency. By promoting green building practices and certifications, we can create healthier living and working spaces that minimize the source and spread of airborne chemicals.

12. Research and Innovation

Continued research and innovation in the field of airborne chemical exposure are vital for developing effective solutions and staying ahead of emerging threats. Scientists, policymakers, and industries must collaborate to explore alternative chemicals and technologies that minimize harm to human health and the environment. By investing in research and supporting innovation, we unlock the potential for disruptive advancements that can revolutionize our approach to mitigating airborne chemical exposure.

As we conclude this chapter on mitigating airborne chemical exposure, it is important to recognize that our collective actions can make a lasting impact. By implementing the practical strategies discussed and embracing innovative approaches, we can safeguard our health, protect our communities, and contribute to a cleaner and healthier future.

Remember, the power to combat the silent spread of airborne chemicals lies in our hands. Let us commit to taking these steps, advocating for change, and inspiring others to join our cause. Together, we can create a world where the menace of airborne chemicals no longer poses a constant threat to our well-being.

Chapter 12: The Role of Governments and Policy

Analyzing the role of governments and policy-making in addressing the menace of airborne chemicals, this chapter discusses the necessary steps to create effective regulations and enforceable standards. In order to combat the potentially devastating impact of these chemicals on human health and the environment, it is imperative for governments to take proactive measures and adopt a comprehensive approach.

Governments around the world play a crucial role in addressing the risks posed by airborne chemicals. By designing and implementing appropriate policies, they can establish a framework that safeguards public health and ensures the protection of our natural resources. The first step in this process is recognizing the urgency of the issue and acknowledging the potential harm caused by these chemicals.

One key aspect of effective government action is the establishment of clear regulations and enforceable standards. Without robust guidelines, it becomes challenging to prevent the release of dangerous chemicals into the air and hold responsible parties accountable for their actions. Governments should focus on conducting thorough research to identify harmful substances and their sources, enabling them to develop specific regulations targeting these sources.

Education and awareness also form an integral part of government initiatives. By providing the public with accurate information about the risks associated with airborne chemicals, governments can empower individuals to make informed decisions regarding their health and well-being. Awareness campaigns, educational programs, and public forums can effectively communicate the potential dangers and encourage individuals to take proactive measures to reduce their exposure.

Collaboration among various stakeholders is essential for the successful implementation of policies related to airborne chemicals. Governments must actively engage with industry leaders, researchers, and environmental organizations to understand diverse perspectives and develop comprehensive strategies. This multifaceted approach ensures that policies take into consideration the practicality, feasibility, and potential impact on different sectors.

Enforcement mechanisms and monitoring systems are vital components of effective policy implementation. Governments need to establish robust regulatory bodies with the authority to enforce standards and guidelines, ensuring compliance across industries. Regular inspections, stringent penalties for non-compliance, and continuous monitoring of air quality are critical to maintaining accountability and preventing the silent spread of airborne chemicals.

Investing in research and innovation is yet another crucial aspect of government involvement in tackling airborne chemical threats. Governments should allocate resources to support scientific studies that investigate the impacts of these chemicals on human health and the environment, as well as explore alternative solutions and technologies. By encouraging research and innovation, governments can foster the development of safer practices, technologies, and materials, which ultimately diminish the risk posed by airborne chemicals.

Moreover, international collaboration and knowledge-sharing must be prioritized. Airborne chemicals are not limited by borders, and their impact can extend far beyond national boundaries. Governments should actively participate in international forums, sharing their experiences, best practices, and expertise to develop a global response to this pervasive and silent spread. Cooperation among nations enables the establishment of harmonized standards and facilitates the exchange of technologies, further fortifying the collective approach against this menace.

In conclusion, governments play a critical role in addressing the menace of airborne chemicals. By formulating effective regulations, raising public awareness, fostering collaboration, enforcing standards, investing in research, and promoting international cooperation, governments can ensure a safer, healthier future for their citizens. It is imperative that governments prioritize the protection of public health and the environment, addressing this silent spread with unwavering determination and vigilance.

Airborne chemicals pose a significant threat to human health and the environment, necessitating the involvement of governments and the implementation of robust policies. In the second half of this chapter, we will delve further into the essential aspects of government action in addressing this menace. From fostering research and innovation to promoting transparency and accountability, governments hold the key to mitigating the risks associated with airborne chemicals.

To effectively combat the spread of airborne chemicals, governments must prioritize the allocation of resources towards research and innovation. By investing in scientific studies and technological advancements, policymakers can gain a deeper understanding of the impacts of these chemicals and identify alternative solutions. By encouraging collaboration between scientists, industry experts, and policymakers, governments can accelerate the development of safer practices, technologies, and materials. By fostering innovation, governments can significantly diminish the potential risks posed by airborne chemicals and ultimately protect public health and the environment.

Transparency and accountability are crucial elements of government action in addressing the menace of airborne chemicals. Governments should establish clear communication channels to provide accurate information about the risks associated with these chemicals. This includes regularly sharing research findings, health warnings, and updates on regulatory measures. Transparent communication builds trust and empowers individuals to make informed decisions regarding their exposure to airborne chemicals.

In addition to transparency, it is vital for governments to establish enforcement mechanisms and monitoring systems. Regulatory bodies should have the authority to enforce standards and guidelines, conduct regular inspections, and impose stringent penalties for non-compliance. Effective monitoring of air quality is also paramount, as it helps identify sources of pollution and assess the efficiency of implemented measures. By enforcing regulations and holding responsible parties accountable, governments can ensure the prevention of the silent spread of airborne chemicals.

Public participation and engagement are integral to successful policy implementation. Governments should actively involve citizens, communities, and organizations in decision-making processes related to airborne chemicals. By creating platforms for public input, such as town hall meetings and public forums, governments can incorporate diverse perspectives, concerns, and knowledge into policy development and implementation. This participatory approach enhances the effectiveness and legitimacy of policies, as it ensures that they address the specific needs and demands of the population.

Furthermore, international collaboration and knowledge-sharing are crucial in tackling the global issue of airborne chemicals. Governments must actively participate in international forums and initiatives to exchange experiences, best practices, and expertise. Through this collaboration, governments can develop harmonized standards that transcend borders and facilitate the exchange of technologies and methodologies. By fostering global cooperation, governments can strengthen the collective response to the silent spread of airborne chemicals, effectively protecting public health and the environment on a global scale.

It is crucial for governments to continuously reassess and adapt their policies to keep pace with evolving scientific knowledge and technological advancements. As new information emerges about the impacts and sources of airborne chemicals, governments must be prepared to adjust regulations and enforcement measures accordingly. Regular monitoring and evaluation of policies are necessary to ensure their effectiveness and identify areas for improvement. By prioritizing the continuous improvement of policies, governments can remain at the forefront of combating the menace of airborne chemicals.

In conclusion, governments play a pivotal role in addressing the menace of airborne chemicals through their formulation and implementation of effective policies. By investing in research and innovation, promoting transparency and accountability, facilitating public participation, and fostering international collaboration, governments can effectively combat the threats posed by airborne chemicals. Through unwavering determination and vigilance, governments can safeguard public health and the environment, ensuring a safer and healthier future for their citizens. The fight against airborne chemicals requires the collective effort of governments worldwide, and it is imperative that we all work together to tackle this silent spread.

Chapter 13: The Power of Advocacy: Voices for Change

In a world where the air we breathe is under constant threat from airborne chemicals, the power of advocacy and grassroots movements cannot be underestimated. These voices for change play a vital role in shedding light on the menace and working towards a safer and healthier environment for all. In this chapter, we highlight the inspiring individuals and organizations that are spearheading the fight against airborne chemical threats.

One such organization is Breathe Easy, a grassroots movement that originated from a small community in Anytown, USA. What started as a concerned group of residents grew into a formidable force advocating for cleaner air. They understood that to bring about change, they needed to come together and demand action.

Breathe Easy quickly captured the attention of local media, which led to increased awareness among the wider population. Their activism did not stop there. They organized rallies and public forums, inviting experts to discuss the effects of airborne chemicals on health and the environment. Through their relentless efforts, they successfully pushed for stricter regulations on industrial emissions, leading to cleaner air and improved lives for their community.

Advocacy can also take on an individual form, as exemplified by Dr. Catherine Evans. Recognized as a leading expert on airborne toxins and their impact on human health, Dr. Evans has dedicated her professional life to advocating for change. Her research and scientific expertise have provided a strong foundation for her advocacy work.

Through her highly acclaimed book, "Toxic Skies," Dr. Evans educates the public about the dangers of airborne chemicals and offers practical solutions to the problem. Moreover, she tirelessly travels the globe, speaking at conferences and engaging with communities affected by toxic air. Dr. Evans believes that knowledge empowers individuals and creates a ripple effect that can lead to a broader societal change.

Another inspiring advocate is Marcus Thompson, a passionate environmentalist who experienced firsthand the devastating consequences of airborne chemical pollution. Growing up in a heavily industrialized area, Marcus witnessed the toll it took on the health of his family and neighbors. Determined to protect future generations, Marcus founded Clean Air Now, an organization dedicated to raising awareness and advocating for stricter regulations on polluting industries.

Through public campaigns, Marcus and his team have successfully mobilized thousands of supporters. Their media-savvy approach has helped them gain attention from influential figures and policy-makers, ultimately leading to significant policy changes. Today, Clean Air Now is regarded as a leading voice in the fight against airborne chemical threats, making a tangible difference in the lives of those affected.

These are just a few examples of the many inspiring individuals and organizations fighting to combat the silent spread of airborne chemicals. Their tireless efforts and unwavering determination remind us of the power of advocacy and grassroots movements. They serve as beacons of hope, proving that change is possible when voices come together in pursuit of a common goal.

As the movement grows, their influence extends beyond local communities, reaching neighboring towns, cities, and even countries. The collective impact of these voices for change is undeniable, making governments and industries take notice. Slowly but surely, they are forcing a shift towards cleaner production practices and more stringent regulations.

However, the battle is far from over. As we will explore in the latter half of this chapter, while progress has been made, there are still significant challenges ahead. To truly unravel the menace of airborne chemicals, we must continue amplifying the voices advocating for change and supporting their efforts. By doing so, we can create a future where the air we breathe is free from harmful substances, ensuring a healthier world for generations to come.As we delve deeper into the second half of this chapter, it becomes evident that the fight against airborne chemical threats is an ongoing battle. While progress has been made through the efforts of inspiring advocates like Breathe Easy, Dr. Catherine Evans, and Marcus Thompson, there are still significant challenges that lie ahead.

One of the challenges faced by advocates is the resistance from industries and policymakers who may be hesitant to implement stricter regulations. The economic implications and potential disruptions in established practices often create roadblocks. However, by sharing scientific research and raising public awareness, advocates can dismantle the misconceptions surrounding the issue and build a stronger case for change.

Advocates like Breathe Easy understand the importance of gathering scientific evidence to strengthen their cause. They form collaborations with researchers and scientists to conduct studies that investigate the effects of airborne chemicals on health and the environment. These studies provide tangible proof, which becomes the foundation for demanding stricter regulations and cleaner production practices.

Furthermore, technology and innovation play a significant role in driving change. The development of air quality monitoring devices allows communities to track pollutant levels and gather data that supports their advocacy efforts. By empowering individuals with this knowledge, advocates can mobilize communities, generate support, and put pressure on the relevant authorities to take action.

In addition to scientific evidence and technological advancements, storytelling is a powerful tool for advocacy. Sharing personal experiences, like Marcus Thompson did with Clean Air Now, can evoke emotions and compassion, leading to a stronger call for action. Personal stories humanize the issue of airborne chemical threats, allowing the public to connect and empathize with the individuals affected.

These stories amplify the voices for change and strengthen the collective resolve towards finding solutions.

Another challenge faced by advocates is the need for international cooperation and collaboration. Airborne chemical threats do not respect borders, and their impact extends far beyond local communities. To truly unravel this menace, advocates must work together across countries, sharing knowledge, resources, and strategies. This includes sharing successful case studies, lessons learned, and best practices in combating airborne chemical threats.

Global conferences and forums provide platforms for experts, organizations, and advocates to come together, share their experiences, and collaborate on finding effective solutions. As a result, international agreements and collaborations have paved the way for the development of global frameworks to address airborne chemical threats. By unifying efforts, advocates increase their influence and maximize the impact they can make.

It is essential to recognize that while advocacy efforts are crucial, they must be complemented by government action. Advocates must push policymakers to prioritize the issue of airborne chemical threats and enact or strengthen legislation that protects the air we breathe. By engaging in dialogue, developing relationships, and presenting well-researched proposals, advocates can shape policy decisions and foster a more supportive and conducive environment for change.

Furthermore, advocating for change at the legislative level involves monitoring, assessment, and enforcement of regulations. Advocates must remain vigilant to ensure that the regulations put in place are followed and enforced effectively, leading to tangible improvements in air quality. This ongoing monitoring helps identify areas where further action is needed and provides evidence for continuous advocacy efforts.

As the chapter draws to a close, it is clear that the power of advocacy and grassroots movements is not to be underestimated. Through their unwavering determination and tireless efforts, individuals and organizations have successfully shed light on the menace of airborne chemical threats. They have inspired communities, influenced policymakers, and initiated significant changes towards a safer and healthier environment.

However, the battle is far from over. The collective efforts of advocates have begun to make a difference, shaping the discourse around airborne chemical threats and forcing governments and industries to take notice. But sustaining this momentum and driving further change requires continued amplification of voices advocating for change and support from individuals, communities, and decision-makers.

Together, we can create a future where the air we breathe is free from harmful substances. By joining hands and persisting in our advocacy, we can ensure the well-being of current and future generations, forging a path towards a healthier and more sustainable world.

Chapter 14: From Awareness to Action: Individual Responsibility

In a world where the very air we breathe is saturated with unseen chemicals, taking back control of our personal space becomes an imperative. The significance of understanding the role each one of us plays in reducing airborne chemical exposure cannot be overstated. It is only by embracing individual responsibility that we can pave the way for a healthier, safer future for ourselves and future generations.

As you embark on this journey towards a less chemical-laden environment, it is important to recognize the power of everyday choices and habits. From the products we select for our homes to the lifestyle choices we make, each decision carries the potential to contribute to a significant reduction in airborne chemical exposure.

At the heart of this transformation lies the conscious decision to prioritize natural, eco-friendly alternatives in our daily lives. Let's start by considering our household cleaning products. Many conventional cleaning agents contain a cocktail of chemicals that can harm not only our respiratory systems but also the environment. By switching to greener alternatives, such as vinegar or baking soda, we can effectively eliminate exposure to harmful airborne pollutants without sacrificing cleanliness.

Furthermore, our homes are not the only spaces where we can take control of the air we breathe. When it comes to personal care products, there exists a vast array of options that are free from harmful chemicals like parabens, phthalates, and sulfates. By incorporating such products into our daily routines, we not only protect our own well-being but also set a positive example for others to follow.

In addition to conscious product choices, our habits and behaviors can play a crucial role in reducing airborne chemical exposure. Let's consider the air quality within our own homes. Opening windows regularly to allow for proper ventilation is a simple yet effective measure to mitigate the accumulation of indoor pollutants. Alongside this, it is important to regularly clean and maintain air filters and vents to ensure optimal functioning.

Taking individual responsibility also extends to making thoughtful choices when it comes to our consumption patterns. In a world inundated with synthetic fragrances, it is essential to be mindful of the products we bring into our homes. Many scented items, such as air fresheners and candles, release harmful volatile organic compounds (VOCs) into the air. Opting for natural alternatives, like essential oil diffusers, not only enhances our well-being but also minimizes exposure to these harmful substances.

As the consciousness of our individual choices deepens, it becomes evident that reducing airborne chemical exposure is not only a matter of personal health but also a collective responsibility. By sharing our knowledge and experiences with others, we can inspire and educate those around us, creating a ripple effect of positive change. Whether it be through conversations with friends and family, involvement in community initiatives, or even using social media platforms, each effort counts.

Awareness is the first step, and action is the driving force that moves us closer to a future where airborne chemical exposure becomes a thing of the past. By embracing our individual responsibility and making conscious choices in our daily lives, we are taking charge of our own well-being and contributing to the well-being of our planet.

As this chapter draws to an end, we have only begun to scratch the surface of the everyday choices and habits that can reduce airborne chemical exposure. In the second half of this chapter, we will delve deeper into the topic, exploring proactive measures from a broader perspective. The journey towards a healthier, less toxic environment continues, and the impact of our individual actions can be far-reaching. Join us in the next half of this chapter as we uncover more secrets to navigating the path from awareness to action.In our journey towards reducing airborne chemical exposure, embracing individual responsibility is only the first step. As we delve deeper into proactive measures, we must broaden our perspective to include not only our immediate surroundings but also the larger environment in which we live. By understanding the interconnectedness of all living beings and ecosystems, we can make informed choices that will have a far-reaching impact.

One crucial aspect of our environment that often goes unnoticed is outdoor air quality. While we may have control over the air within our homes, the air we breathe outside is equally important. By advocating for and supporting policies and practices that prioritize clean air, such as stricter emissions standards for vehicles and industrial facilities, we can contribute to a healthier atmosphere for everyone.

Furthermore, reducing our reliance on transportation that contributes to air pollution is another way to take action. Whenever possible, opting for eco-friendly modes of transportation, such as walking or cycling, can significantly reduce both our personal exposure to air pollutants and our carbon footprint. Carpooling or using public transportation are also effective ways to minimize emissions.

Additionally, being mindful of the products we use and dispose of is crucial in our quest for cleaner air. Chemicals from everyday items often find their way into the air and water through improper disposal methods. By responsibly disposing of hazardous materials, recycling whenever possible, and supporting businesses that prioritize sustainability, we actively contribute to a cleaner environment.

Understanding the importance of green spaces in our communities is another aspect of individual responsibility. Trees, plants, and other natural elements not only provide us with clean air but also offer numerous other benefits, such as reducing the urban heat island effect and enhancing overall well-being. Supporting initiatives that aim to preserve and expand green spaces, whether through volunteering or advocacy, is an effective way to contribute to air quality improvement.

Education and awareness play a significant role in inspiring action within our communities. By sharing our knowledge and experiences, we can engage others in the importance of reducing airborne chemical exposure. Hosting workshops, participating in local events, or even starting a conversation with friends and neighbors can spread awareness and encourage proactive change.

The power of collective action should not be underestimated. Engaging with local and national organizations that focus on environmental protection can amplify our individual efforts. Joining forces with like-minded individuals can bring about policy changes that prioritizes clean air and reduce the use of harmful chemicals in various industries.

Furthermore, as consumers, we have the power to influence markets and demand safer alternatives. By supporting companies that prioritize eco-friendly practices and sustainable production methods, we send a clear message that our health and the health of the planet are paramount. Familiarizing ourselves with eco-labeling systems and discerning between truly sustainable products and those that use greenwashing tactics empowers us to make informed purchasing decisions.

Finally, social media platforms can be a powerful tool in spreading awareness and promoting individual responsibility. Utilizing these platforms to share informative content, highlight success stories, and connect with others passionate about creating a healthier environment can exponentially increase our impact.

As we reach the conclusion of this chapter, our journey from awareness to action is just beginning. The choices we make every day, both big and small, have the potential to reduce airborne chemical exposure and contribute to a healthier future for ourselves and the generations to come. By embracing our individual responsibility, advocating for systemic change, and fostering collective action, we can create a ripple effect of positive change that transcends borders and transforms our world.

In closing, I invite you, the reader, to reflect on the knowledge you have gained and the actions you can take in your own life. Remember that the responsibility lies with each one of us to make a difference. Together, let us continue our commitment to reducing airborne chemical exposure and ensure a cleaner, healthier future for all.

Chapter 15: Innovations and Breakthroughs in Air Quality

As the threat of airborne chemicals continues to loom large, scientists and researchers worldwide have been tirelessly exploring innovative solutions and breakthroughs in air quality management. In this chapter, we delve into the realm of cutting-edge technologies and approaches that promise to revolutionize the way we address air pollution. From advanced filtration systems to gene editing techniques, the possibilities are endless.

One remarkable innovation that holds great promise in the field of air quality management is the development of nanotechnology-based air filters. Traditional air filters have long been utilized to trap particulate matter and pollutants, but nanotechnology takes this concept to a whole new level. Nanofilters, designed with microscopic precision, are capable of capturing even the tiniest particles, including harmful chemicals and ultrafine particulate matter. These filters utilize nanostructured materials, such as carbon nanotubes and graphene, to create a highly efficient and selective filtration system. The integration of nanofilters into existing ventilation systems has the potential to significantly improve indoor air quality, safeguarding our health and well-being.

In addition to nanotechnology-based filters, another exciting breakthrough in air quality management lies in the field of biotechnology. Scientists have been exploring the potential of genetically engineered microorganisms to combat air pollution. By harnessing the power of synthetic biology, researchers have successfully engineered bacteria that can metabolize and break down harmful chemicals present in the air. Through genetic modifications, these microorganisms can be optimized to target specific pollutants, offering a tailored approach to air quality control. This emerging field of bioaugmentation holds immense promise, and ongoing research and development efforts are expected to yield groundbreaking results in the near future.

Furthermore, artificial intelligence (AI) has emerged as a powerful tool in the fight against airborne chemicals. By leveraging machine learning algorithms and data analytics, AI systems can continuously monitor air quality in real-time, enabling prompt intervention and mitigation measures. These intelligent systems can analyze vast amounts of data collected from sensors positioned throughout various locations, enabling early detection of pollutants and the ability to predict potential air quality hazards. This proactive approach towards air quality management empowers decision-makers with valuable insights and allows for timely implementation of targeted solutions.

Moreover, advancements in sensor technology have paved the way for the development of portable air quality monitoring devices. These compact yet sophisticated sensors are capable of detecting and quantifying various pollutants in real-time. With the ability to instantly assess the quality of the air we breathe, individuals can make informed decisions to protect their health and well-being. Such devices not only raise awareness about personal exposure to pollutants but also contribute to a collective effort in combating air pollution on a larger scale.

While the first half of this chapter sheds light on the remarkable innovations and breakthroughs in air quality management, there is still much more to explore. In the upcoming pages, we will delve into novel solutions and groundbreaking research that push the boundaries of air quality control. The challenges we face are immense, but with each scientific advancement, we move closer to unraveling the menace of airborne chemicals.

As the story continues in the second half of this chapter, we invite you to embark on a journey through the fascinating world of air quality management. Brace yourself for surprising revelations, unimaginable possibilities, and the relentless pursuit of a cleaner and healthier future. Stay tuned for the next installment, where we delve deeper into the cutting-edge technologies that hold the potential to reshape our relationship with the air we breathe. The silent spread of airborne chemicals may seem insurmountable, but through innovation and scientific breakthroughs, we are laying the groundwork for a safer and more sustainable world.In the second half of this chapter, we delve deeper into the extraordinary advancements and ongoing research that are reshaping the field of air quality management. The relentless pursuit of a cleaner and healthier future continues, with scientists and researchers breaking new ground to combat the silent spread of airborne chemicals.

One groundbreaking area of research that shows tremendous potential is the application of photonics in air quality control. Photonics, the science of generating, controlling, and detecting photons (particles of light), opens up a whole new frontier in the development of sensors and monitoring devices. Researchers are exploring the use of laser-based technology to detect and identify airborne pollutants with incredible precision and sensitivity. These advanced systems can analyze the molecular composition of air samples, enabling rapid identification of harmful chemicals and accurate quantification of their concentrations. This level of accuracy and specificity allows for targeted interventions and effective mitigation strategies.

Furthermore, the field of air quality management is witnessing remarkable progress in the development of eco-friendly building materials. Traditional construction materials often release volatile organic compounds (VOCs) and other harmful pollutants into the air, significantly affecting indoor air quality. However, innovative solutions are emerging, such as low-emission paints, adhesives, and sealants. These environmentally friendly materials are specially formulated to minimize the release of pollutants, ensuring healthier indoor environments.

In addition to eco-friendly materials, sustainable architectural designs are gaining traction in the quest for better air quality. Building structures that incorporate natural ventilation systems and green infrastructure can help reduce the reliance on energy-intensive mechanical ventilation while improving outdoor and indoor air circulation. Techniques such as the use of vertical gardens, rooftop green spaces, and natural shading systems contribute to the purification and filtration of outdoor air, ultimately benefiting the overall air quality in urban areas.

Another area of exploration with immense potential lies in the realm of biomimicry, drawing inspiration from nature to develop effective air quality management strategies. Researchers are studying how organisms in natural ecosystems remove and filter pollutants from the air, with the aim of replicating these processes in human-made systems. By mimicking the intricate mechanisms present in plants, animals, and microorganisms, scientists hope to design artificial systems capable of trapping and neutralizing harmful airborne chemicals.

Moreover, the integration of advanced data analytics into air quality management systems is revolutionizing the way we approach pollution control. By harnessing the power of big data and sophisticated algorithms, these systems can generate real-time air quality models and predictions, enabling timely intervention and more efficient resource allocation. This data-driven approach supports evidence-based decision-making and empowers policymakers and authorities to implement targeted measures for pollution reduction.

As research in air quality management advances, so too does our understanding of the synergistic effects of different pollutants. Scientists are investigating how different chemicals interact and combine to create new and potentially more harmful compounds. This knowledge is essential for designing effective mitigation strategies and developing appropriate regulations to tackle complex air pollution challenges comprehensively.

In the quest for cleaner air, it is essential to engage and educate the public about the importance of individual actions in safeguarding air quality. Community-driven initiatives, public awareness campaigns, and educational programs are vital in fostering a collective sense of responsibility and promoting sustainable practices. By empowering individuals to make informed choices, we can create a ripple effect and promote long-term behavioral change.

In conclusion, the second half of this chapter has provided a glimpse into the remarkable advances and ongoing research in air quality management. From photonics and eco-friendly materials to biomimicry and data analytics, these innovative approaches are transforming our understanding and control of airborne chemicals. As each discovery pushes the boundaries of our knowledge, we inch closer to unraveling the menace of air pollution. The journey towards a cleaner, healthier future requires continued commitment, collaboration, and the unwavering pursuit of scientific breakthroughs. Together, we can overcome the challenges and create a sustainable world in which the air we breathe is safe and pure. Stay tuned for the next installment, where we explore promising avenues for air quality management.

Chapter 16: Addressing Airborne Chemicals in the Workplace

Occupational hazards pose a significant threat to the well-being of employees across various industries. Among these hazards, airborne chemicals stand out as a silent menace that can have detrimental effects on both physical health and overall productivity. In this chapter, we explore the steps that employers and employees can take to create safer and healthier work environments.

To begin, it is crucial to understand the types of airborne chemicals that may be present in workplaces. These can range from volatile organic compounds (VOCs), such as benzene and formaldehyde, to particulate matter and gases emitted from machinery and equipment. While some of these chemicals are an inevitable part of certain industries, effective measures can be adopted to minimize exposure and mitigate potential risks.

The first line of defense lies in comprehensive risk assessments conducted by employers. By identifying the specific airborne chemicals present in the workplace, employers can develop targeted strategies for their mitigation. This includes implementing engineering controls, which aim to minimize or eliminate exposures at their source. Examples of engineering controls may include adequate ventilation systems, air filters, and isolation enclosures for high-risk processes.

Furthermore, personal protective equipment (PPE) becomes vital in safeguarding employees from airborne chemical hazards. Properly fitted respirators, protective clothing, and safety goggles can significantly reduce the potential for inhalation or skin contact with harmful substances. Employers must provide and ensure the proper use of PPE, as well as conduct regular training on its correct application.

In addition to the responsibilities of employers, employees themselves play a crucial role in maintaining a safe work environment. Familiarizing oneself with the chemicals encountered in the workplace and their associated risks is fundamental. This knowledge allows employees to recognize potential hazards and take appropriate precautions. Participating in training programs that cover chemical safety protocols and emergency response procedures further empowers employees to protect themselves and their colleagues.

Collaboration and communication between employers and employees form the cornerstone of effective measures to address airborne chemicals. Employers should establish clear lines of communication where employees are encouraged to report any concerns or potential hazards promptly. Creating a culture that values open dialogue and encourages proactive participation enables employers to swiftly address emerging risks.

Regular monitoring of air quality within the workplace is critical for ensuring that established control measures are effective. This can be achieved through air sampling and analysis, conducted by qualified professionals. Monitoring provides valuable data on the presence and concentration of airborne chemicals, aiding in ongoing risk assessments and the refinement of control strategies.

Efforts to address airborne chemicals in the workplace should not be confined solely to reactive measures. Proactive steps, such as substituting hazardous substances with less harmful alternatives, should be considered whenever feasible. Employers should explore greener technologies and materials, reducing the reliance on harmful chemicals and promoting a sustainable work environment.

While numerous regulations and guidelines exist to govern occupational health and safety, compliance alone is not enough. Employers must go beyond the minimum requirements, taking responsibility for the overall well-being of their workforce. Prioritizing the reduction of airborne chemical hazards requires a commitment to ongoing education, employee engagement, and continuous improvement.

As we have explored in this first half of the chapter, addressing airborne chemicals in the workplace demands a collaborative effort between employers and employees. Comprehensive risk assessments, engineering controls, proper use of personal protective equipment, and proactive measures contribute to a safer environment. However, the journey towards creating healthier workspaces does not end here. In the second half of this chapter, we will delve deeper into effective training programs, emergency response procedures, and emerging technologies that can further enhance workplace safety. Stay tuned for the continuation of this crucial discussion.In the second half of this chapter, we will delve deeper into effective training programs, emergency response procedures, and emerging technologies that can further enhance workplace safety. These components are essential in equipping employees with the knowledge and skills necessary to identify and address airborne chemical hazards in an efficient and timely manner.

One of the foundational aspects of promoting workplace safety is providing comprehensive training programs to all employees. These programs should cover various topics, including the identification of different airborne chemicals, understanding their potential health effects, and the appropriate measures to prevent exposure. By educating employees about the specific chemicals present in their workplace and the associated risks, employers empower them to make informed decisions and take necessary precautions.

While initial training is crucial, there is also a need for ongoing education to ensure that employees remain up-to-date with the latest advancements and best practices in chemical safety. Regular refresher courses and workshops can reinforce knowledge and serve as a platform for addressing any questions or concerns. Furthermore, staying informed about emerging technologies and alternative materials can help employees and employers explore innovative ways to further mitigate the risks associated with airborne chemicals.

Alongside comprehensive training, clear emergency response procedures are vital for swift action in the event of an airborne chemical incident. Establishing a well-defined protocol ensures that employees know what steps to take to protect themselves and others in the event of a chemical release or exposure. This protocol should include guidelines for evacuations, communication channels to alert relevant personnel, and instructions for administering first aid when necessary. Regular drills and simulations can also enhance preparedness, allowing employees to practice their roles and become familiar with emergency response procedures.

Advancements in technology offer new avenues for improving workplace safety related to airborne chemicals. For instance, the development of air quality monitoring systems provides real-time data on the concentration and presence of harmful chemicals within the workplace. These systems can alert both employees and employers to sudden changes in air quality, facilitating swift responses and the implementation of appropriate control measures. Furthermore, wearable sensors that detect and measure chemical exposures offer employees an enhanced level of personal protection and continuous monitoring, leading to prompt action when necessary.

In recent years, the concept of "green chemistry" has gained traction as a way to reduce the use of hazardous chemicals and promote sustainability in the workplace. Employers should explore and invest in greener technologies and materials to minimize or eliminate the need for harmful chemicals. This shift can contribute not only to a safer work environment but also to a more environmentally friendly one, aligning with the growing global focus on sustainability and corporate social responsibility.

To ensure the effectiveness of all these measures, regular evaluation and revision of control strategies are necessary. Periodic audits and assessments should be conducted to identify any gaps or weaknesses in existing systems. These evaluations can help employers identify areas for improvement, update procedures based on new regulations or technologies, and ensure ongoing compliance with safety standards.

In conclusion, addressing airborne chemicals in the workplace requires a multifaceted approach that encompasses training programs, emergency response procedures, and the integration of emerging technologies. By prioritizing employee education, establishing clear protocols, and capitalizing on advancements in air quality monitoring and green chemistry, employers can create work environments that are safer, healthier, and more sustainable.

Through the collaborative efforts of employers and employees, coupled with a commitment to ongoing education and continuous improvement, the menace of airborne chemicals can be successfully unraveled. By prioritizing the well-being of the workforce, organizations can foster a culture of safety, protection, and productivity. Let us continue our journey towards healthier workspaces, guided by the principles and practices outlined in this chapter.

Chapter 17: Children and Airborne Chemicals: Protecting the Vulnerable

Introduction:

Airborne chemicals pose a significant threat to the health and well-being of children, who are particularly vulnerable to their harmful effects. As their bodies and immune systems are still developing, children face unique risks when exposed to these invisible menaces in the air. This chapter delves into the exploration of these risks and aims to unravel strategies to safeguard the health of our young ones from the dangers of airborne chemicals.

Understanding the Vulnerabilities:

Children, with their smaller size and faster metabolism, breathe in more air and absorb greater amounts of pollutants compared to adults. Their developing bodies also undergo critical phases of growth and development that can be significantly disrupted by exposure to harmful airborne chemicals. Additionally, children often engage in activities where they spend more time outdoors, increasing their potential exposure to air pollutants.

Impact on Physical Health:

Exposure to airborne chemicals can lead to short-term and long-term health consequences for children. Short-term effects can manifest as respiratory issues, such as asthma exacerbation, bronchitis, or allergic reactions. Prolonged exposure may result in chronic respiratory conditions that persist into adulthood, compromising lung function and overall well-being. Chemicals present in the air have also been linked to neurodevelopmental disorders, cardiovascular diseases, and even cancer in children, posing serious long-term health risks.

Psychosocial Impact:

Beyond their physical well-being, children also suffer psychological and emotional consequences due to the menace of airborne chemicals. Constant exposure to polluted air can lead to increased anxiety, stress, and even depression among children. This can further impact their ability to concentrate, learn, and thrive academically and socially. The recognition of this psychosocial impact highlights the urgency and importance of protecting children from airborne chemicals.

Strategies for Safeguarding the Young Ones:

Governments, policymakers, communities, and parents share the responsibility of implementing strategies that protect children from the harmful effects of airborne chemicals. Here are some crucial steps that can be taken to ensure the safety of our young ones:

1. Strengthen Environmental Regulations:

Governments should prioritize the implementation and enforcement of stringent regulations that limit the release of hazardous pollutants into the air. This includes both industrial and transportation sectors, encouraging the adoption of cleaner technologies and promoting sustainable practices.

2. Indoor Air Quality Management:

Since children spend a significant amount of time indoors, ensuring good indoor air quality is essential. This can be achieved through proper ventilation systems, regular maintenance of heating, ventilation, and air conditioning (HVAC) systems, and minimizing the use of chemical products indoors.

3. Education and Awareness:

Increasing public awareness about the risks posed by airborne chemicals is crucial. Parents, educators, and healthcare professionals should be equipped with the necessary knowledge to identify potential sources of pollutants and take appropriate measures to reduce exposure. Educational campaigns and outreach programs should be conducted to empower the community with preventive actions.

4. Healthy Urban Planning:

Urban areas often face higher levels of air pollution due to industrial activities and dense traffic. Incorporating green spaces, tree planting initiatives, and promoting sustainable transportation options can help reduce pollution levels and provide cleaner air for children to breathe.

5. Engaging Industry:

Collaboration with industries is essential in curbing pollution and protecting children's health.

Encouraging industries to adopt cleaner production methods, invest in pollution control technologies, and reduce their emissions can have a significant impact on improving air quality for children.

Conclusion:

As the primary concern of every society, the protection of children should extend to safeguarding their health from airborne chemicals. By addressing the vulnerabilities and risks faced by children and implementing robust strategies, we can take crucial steps toward ensuring a healthier future for the young ones. As this chapter has highlighted, it is imperative that we acknowledge the unique challenges faced by children, empower communities, and work together to create a safe and healthy environment for their growth and well-being.

Assessing Exposure Risks: Analyzing the factors that contribute to children's exposure to airborne chemicals is essential in formulating effective protective measures. This section delves into the evaluation of exposure risks and explores ways to mitigate them.

1. Identifying High-Risk Environments:

Certain settings pose a higher risk of exposing children to airborne chemicals. Industrial areas, densely populated regions, and proximity to major roadways are examples of environments where pollutant concentrations are typically elevated. By conducting thorough air quality assessments in these areas, authorities can determine the extent of the risks and take targeted actions to reduce exposure for children residing or spending time in these zones.

2. Monitoring Indoor Air Quality:

While outdoor pollutants are a significant concern, studies have shown that indoor air can also harbor hazardous chemicals. Monitoring indoor air quality in homes, schools, and daycare centers is vital to identify potential sources of exposure. Regular testing can involve measuring the levels of volatile organic compounds (VOCs), allergens, and particulate matter (PM), ensuring that indoor environments are safe and healthy for children.

3. Consumer Product Safety:

Children can come into contact with harmful chemicals through various consumer products, including furniture, toys, cleaning agents, and personal care items. Governments and regulatory bodies should implement strict guidelines and testing protocols to ensure the safety of products intended for children's use. This includes monitoring and limiting the presence of toxic substances such as lead, flame retardants, and phthalates in these items.

4. Assessing Occupational Exposures:

Children engaged in certain occupations, such as agriculture or industrial work, may face heightened risks of exposure to harmful chemicals. It is crucial to establish regulations prohibiting child labor in hazardous environments and promote alternatives such as education and skill development opportunities to protect vulnerable young individuals from dangerous occupational exposures.

Preventive Measures:

Acknowledging the risks is only the first step; implementing preventive measures is crucial to protect children from the harmful effects of airborne chemicals. The following strategies outline actionable steps that can be taken to safeguard the health and well-being of our young ones.

1. Enhanced Surveillance Systems: Establishing robust monitoring systems and platforms for reporting health outcomes related to airborne chemical exposures can help in identifying vulnerable communities and implementing targeted interventions. By collecting and analyzing data on healthcare utilization, such as emergency room visits or respiratory illnesses, authorities can identify areas with higher prevalence rates and allocate resources accordingly.

2. Protective Gear and Hygiene Practices: In certain situations where exposure cannot be entirely avoided, providing children with appropriate protective gear, such as face masks or respirators, can help minimize their risk. Additionally, promoting good hygiene practices, such as handwashing after outdoor activities or before meals, can reduce the chances of ingesting or absorbing harmful chemicals inadvertently.

3. Collaborative Efforts: Protecting children from airborne chemicals requires a multidisciplinary approach and collaboration among stakeholders. Governments, non-governmental organizations, healthcare professionals, educators, parents, and communities should join forces to raise awareness, share knowledge, and coordinate efforts in enforcing regulations and implementing protective measures.

4. Research and Innovation: Investing in research and development is essential to advance our understanding of the health effects of airborne chemicals on children. By fostering scientific inquiry, innovative solutions can be developed to reduce pollution sources, improve air quality, and mitigate the risks faced by our young ones. This includes exploring renewable energy alternatives, promoting green technologies, and developing efficient air filtration systems for both indoor and outdoor environments.

5. Empowering Communities: Educating and empowering communities is vital in creating a sense of ownership and responsibility toward protecting children from airborne chemical risks. Public engagement initiatives, community workshops, and educational programs can equip parents, caregivers, and educators with knowledge and skills to reduce exposure risks within their immediate surroundings and advocate for broader changes in air quality management.

Conclusion: Safeguarding the vulnerable population of children from the dangers of airborne chemicals requires a comprehensive and unwavering commitment from all facets of society. By understanding their unique vulnerabilities, implementing stringent regulations, promoting healthy urban planning, and engaging in collaborative efforts, we can work together to create an environment where children can thrive without the detrimental impacts of harmful airborne chemicals. It is our duty to protect the health and well-being of our young ones, ensuring that they inherit a world free from the silent spread and its menace.

Chapter 18: Beyond Airborne Chemicals: Emerging Challenges

As we continue to grapple with the silent spread of airborne chemicals and their detrimental effects on our environment and health, it is crucial that we look ahead and anticipate the emerging challenges that may further exacerbate air pollution. The world around us is constantly evolving and bringing forth new threats that demand our attention and action.

One of the key emerging challenges is the rapid urbanization and population growth that our planet is witnessing. As more and more individuals migrate to cities and urban areas, the demand for resources and energy increases significantly. This surge in demand often translates into heightened levels of air pollution, as industries and transportation systems struggle to keep up with the pace of urban expansion.

Furthermore, the increasing reliance on fossil fuels contributes heavily to the rising air pollution levels. As we strive for progress and economic growth, the combustion of fossil fuels remains a primary energy source for a majority of countries. The emissions from burning coal, oil, and gas release copious amounts of greenhouse gases and other air pollutants, adding to the existing burden on our atmosphere.

Moreover, emerging challenges in transportation pose another threat to air quality. Although strides have been made in developing electric vehicles and promoting public transportation, the sheer number of vehicles on the road continues to escalate. The exhaust emissions from conventional automobiles, particularly in densely populated areas with heavy traffic, release harmful pollutants such as nitrogen oxides and particulate matter. These pollutants not only pose immediate health risks but also contribute to the formation of long-term air pollutants such as ground-level ozone.

In addition to these ongoing challenges, emerging sources of air pollution further complicate the fight against airborne chemicals. One such source is the incense burning and house-hold cleaner industry. Incense burning is a common cultural practice in several regions, releasing volatile organic compounds and fine particulate matter into the air. Similarly, the use of household cleaning products often contains chemical compounds that can be harmful when inhaled. As our understanding of the long-term effects of these sources grows, it becomes evident that they too should be scrutinized and addressed to mitigate their impact on air quality.

While it is essential to recognize and address the immediate threats posed by airborne chemicals, it is equally important to acknowledge the emerging challenges and potential future hazards. By doing so, we can develop comprehensive strategies that encompass a broader range of pollutants and devise long-term solutions that promote sustainable development.

Future chapters in this book will delve deeper into these emerging challenges and explore potential solutions to combat air pollution. As we continue our journey, it is crucial to remember that the fight against airborne chemicals is not limited to a single aspect of pollution but encompasses a broader scope that requires collaborative efforts and continuous innovation. Only by staying vigilant and adapting to the ever-changing landscape can we hope to overcome the menace of airborne chemicals and safeguard our planet for future generations.

As we delve deeper into the challenges posed by airborne chemicals and explore potential solutions to combat air pollution, it becomes evident that safeguarding our planet requires a comprehensive and multidimensional approach. In this second half of Chapter 18, we will further explore emerging challenges and their impact on air quality, as well as discuss innovative strategies to mitigate their effects.

One of the emerging challenges we face is the rapid advancement of technology and its influence on air pollution. While technological developments have undoubtedly improved our lives in various aspects, they have also brought forth new sources of pollution. The widespread use of electronic devices, including smartphones, laptops, and tablets, has resulted in the generation of electronic waste or e-waste. When improperly disposed of, these devices can release toxic substances such as lead, mercury, and brominated flame retardants into the environment, contaminating the air, water, and soil.

To address this challenge, it is crucial to promote responsible electronic waste management practices. Implementing effective recycling programs and encouraging the use of sustainable materials in the production of electronic devices can help minimize the environmental impact of e-waste and decrease air pollution caused by their improper disposal. Furthermore, raising awareness among consumers about the importance of recycling and proper electronic waste disposal is paramount in reducing the negative consequences associated with technological advancements.

Another consequential emerging challenge is the impact of climate change on air quality. The steady rise in global temperatures and the resultant climate patterns disrupt atmospheric conditions, leading to the formation of secondary pollutants such as ground-level ozone. Ozone, a major component of smog, poses significant health risks, particularly to those with respiratory conditions and the elderly. Additionally, changing weather patterns, including more frequent and intense heatwaves, can exacerbate the formation of ozone and other pollutants.

To mitigate the detrimental effects of climate change on air quality, addressing the root cause is imperative. Implementing measures to reduce greenhouse gas emissions, such as transitioning to renewable energy sources and improving energy efficiency, is crucial. Moreover, fostering global cooperation and committing to ambitious emissions reduction targets are essential steps towards combatting climate change and ensuring cleaner air for current and future generations.

Emerging challenges also arise through the use of certain chemicals in various industries and consumer products. Flame retardants, for example, are commonly found in household items such as furniture, textiles, and electronics. These chemicals have been associated with adverse health effects, including respiratory issues and neurodevelopmental disorders. Effective regulation and stricter enforcement of flame retardant usage, coupled with the development and adoption of safer alternatives, are critical actions that can help alleviate their impact on air quality.

Furthermore, the expansion of industrial activities, particularly in developing countries, presents an ongoing challenge. The rapid economic growth observed in many regions is often accompanied by an increase in industrial emissions, resulting in elevated levels of air pollution. To address this challenge, it is essential to establish and enforce comprehensive regulations that govern industrial emissions, promote cleaner production techniques, and encourage the adoption of sustainable technologies. Investing in research and development of low-emission industrial processes can also contribute to reducing the impact of industrial activities on air quality.

In addition to addressing emerging challenges, it is imperative to continuously monitor and evaluate existing control measures to ensure their effectiveness. Robust air quality monitoring systems, combined with data analysis and modeling techniques, play a crucial role in assessing pollution levels, identifying pollution hotspots, and gauging the effectiveness of implemented strategies. Regular evaluation and adaptation of control measures based on scientific evidence and technological advancements are vital to continually improve air quality and protect human health.

As we conclude this chapter, it is evident that the fight against airborne chemicals extends beyond solely addressing their immediate threats. Emerging challenges, such as rapid urbanization, technological advancements, climate change, and the use of certain chemicals, further exacerbate air pollution and demand comprehensive strategies and solutions. By fostering collaboration, innovation, and sustained efforts, we can work towards a future with clean and breathable air for everyone.

Closing the chapter on airborne chemicals, we will continue our exploration of other facets of environmental challenges in the subsequent chapters of this book. Each topic will shed light on a different aspect of our interconnected world, emphasizing the significance of collective action and informed decision-making. Let us embark on this journey, armed with knowledge and determination, to safeguard our planet for present and future generations.

Chapter 19: The Path to a Healthier Future: Policy Recommendations

In order to navigate the complex landscape of airborne chemicals and their detrimental impact on public health, it is imperative that we propose tangible and actionable policy recommendations. These recommendations will form the foundation for a roadmap towards a healthier future, where collective action and comprehensive solutions take center stage.

1. Strengthening Air Quality Standards:

Our first policy recommendation involves the urgent need to strengthen air quality standards. Existing standards have become outdated, failing to address the emerging threats posed by airborne chemicals. By updating and enforcing stringent regulations, we can effectively reduce the presence of harmful substances in the air we breathe. This should include a reassessment of permissible emission thresholds and the incorporation of new, unseen chemical compounds that threaten human health.

2. Implementing Comprehensive Monitoring Systems:

To effectively combat the silent spread of airborne chemicals, it is essential to establish robust monitoring systems. These systems should stretch across residential, industrial, and commercial areas, providing real-time data on air quality. By collecting accurate information on pollutant levels, we can identify hotspots and target interventions accordingly. This data-driven approach will enable policymakers and scientists to make informed decisions, prioritizing areas most urgently in need of intervention.

3. Promoting Green Innovation:

Transitioning towards a healthier future necessitates a focus on green innovation. Policymakers must incentivize industries to adopt sustainable practices, reducing the emissions of harmful chemicals in the manufacturing process. Research and development funding should be directed towards the development of eco-friendly alternatives that pose minimal risks to human health and the environment. By harnessing the power of science and technology, we can achieve a future where industrial growth and public health coexist harmoniously.

4. Empowering Public Awareness and Education:

One of the foremost challenges in combating the silent spread of airborne chemicals lies in the lack of public awareness. Ensuring the success of any policy recommendations requires a concerted effort to educate and empower the public. Implementing awareness campaigns, educational programs, and user-friendly resources will equip individuals with the knowledge to protect themselves and advocate for their communities' well-being. Only through collective action can we catalyze the necessary change at the grassroots level.

5. Enhancing International Cooperation:

Airborne chemicals do not respect geopolitical boundaries, highlighting the need for robust international cooperation. Collaborative efforts should be established to share data, research findings, and best practices. By transcending borders, policymakers and scientists can collectively address global challenges arising from airborne chemicals. This collaboration will promote cohesive policy frameworks, ensuring a unified approach in safeguarding public health worldwide.

As we embark on this journey towards a healthier future, it is crucial to recognize that no single policy recommendation alone can solve the complex issue of airborne chemicals. Rather, a combination of these thoughtful approached is needed. By strengthening air quality standards, implementing comprehensive monitoring systems, promoting green innovation, empowering public awareness, and enhancing international cooperation, we forge a path towards a future where the silent spread of airborne chemicals no longer poses a menace to humanity.

As we delve deeper into the path towards a healthier future, it is evident that comprehensive solutions and collective action must remain at the forefront of our endeavors. To complete the roadmap addressing the menace of airborne chemicals, let us now turn our attention to the remaining policy recommendations that will play a vital role in mitigating the impact of these harmful substances.

6. Fostering Collaborative Research Initiatives:

Advancing our understanding of the numerous airborne chemicals and their effects on public health requires concerted research efforts. Governments, academic institutions, and private organizations should collaborate to fund research projects aimed at identifying and assessing the presence of these chemicals. By pooling resources, expertise, and data, we can accelerate the development of strategies to combat the silent spread of airborne chemicals.

Such collaborative research initiatives will involve characterizing and monitoring emerging chemical compounds, identifying their sources, and elucidating their detrimental effects on human health. Transparent sharing of research findings and methodologies will ensure that efforts are not duplicated and that scientific advancements are maximized for the benefit of public well-being worldwide.

7. Strengthening Regulatory Frameworks:

Existing regulations often lag behind the rapid evolution of technology and the emergence of new airborne chemicals. To counter this, we must enhance national and international regulatory frameworks for the detection, assessment, and control of these substances. Industries should be held accountable for their emissions and required to abide by stricter standards that reflect the latest scientific knowledge regarding airborne toxins.

Moreover, regulatory bodies should be allocated greater resources and authority to enforce compliance, ensuring that companies strictly adhere to the stringent air quality standards. By promoting transparency, accountability, and rigorous enforcement, we can make significant progress in reducing the health risks associated with airborne chemicals.

8. Investing in Sustainable Infrastructure:

Our built environment plays a critical role in the distribution and accumulation of airborne chemicals. To minimize exposure, sustainable urban planning and infrastructure development must be prioritized. Green spaces, such as parks and urban forests, should be integrated into city planning to enhance air quality and create natural buffers against pollution.

Additionally, the construction sector must adopt sustainable practices that prioritize the use of non-toxic materials and energy-efficient designs. By investing in sustainable infrastructure, we can mitigate the release of harmful chemicals and create healthier living environments for current and future generations.

9. Supporting Vulnerable Communities:

The impact of airborne chemicals is often disproportionately borne by marginalized and vulnerable communities. Environmental justice must be a central tenet of our policy recommendations, addressing the disparities in exposure and ensuring that these communities have the necessary resources to protect themselves.

This includes providing financial assistance for the installation of air filtration systems for low-income households and schools located in areas with high pollution levels. Furthermore, initiatives should be launched to empower communities through education and capacity-building programs, enabling them to advocate for their rights and implement localized solutions.

10. Strengthening Global Governance:

Uniting nations under a common goal of mitigating the harm caused by airborne chemicals necessitates strengthening global governance structures. International organizations, such as the World Health Organization (WHO), should play an active role in coordinating efforts, setting global standards, and promoting accountability.

Robust international agreements and protocols should be established to regulate the production, transportation, and disposal of chemicals with harmful properties. Cooperation among nations will ensure a harmonized approach to combating the menace of airborne chemicals, transcending geopolitical boundaries and safeguarding the health of all nations.

In combining these additional policy recommendations with the ones presented earlier, we establish a holistic framework capable of addressing the complex issue of airborne chemicals. Strengthening air quality standards, implementing comprehensive monitoring systems, promoting green innovation, empowering public awareness, enhancing international cooperation, fostering collaborative research initiatives, strengthening regulatory frameworks, investing in sustainable infrastructure, supporting vulnerable communities, and strengthening global governance collectively pave the road to a healthier future.

Let us remember that overcoming the challenges posed by airborne chemicals requires a multi-faceted approach. It demands the collaboration of governments, industries, scientific communities, and individuals alike. Striving for a healthier future means acknowledging the significance of these policy recommendations and taking action upon them without delay.

In conclusion, as we continue our journey towards healthier air and improved public health, let us remain resolute in our pursuit of a world where the silent spread of airborne chemicals no longer poses a menace to humanity. By implementing these policy recommendations, we can usher in an era of cleaner, safer air for generations to come, protecting the well-being of all and leaving a healthier world as our legacy.

Chapter 20: Empowering Communities: Taking a Stand

In the face of a growing threat from airborne chemicals, it is essential for communities to rise up, unite, and take a stand. Throughout this book, we have explored the silent spread of these invisible menaces, the impact they can have on our health and the environment, and the urgent need for change. Now, in this concluding chapter, we will delve into the power of communities and individuals, and how they can become the driving force behind tackling these airborne chemical threats.

Communities are the backbone of society, and when they come together with a common purpose, they become an unstoppable force. Each city, town, and neighborhood can take the initiative to address and combat the dangers of airborne chemicals that pervade their surroundings. The key lies in empowering individuals to become informed, proactive, and catalysts for positive change.

Education and awareness form the foundation of any successful movement. By educating ourselves and our neighbors about the hazards posed by airborne chemicals, we can create a ripple effect of knowledge throughout our communities. This chapter aims to equip readers with the information they need to become advocates in their own right, arming them with the tools required to lead the charge against this silent menace.

The first step in empowering communities is to establish strong networks, where concerned individuals can come together, share experiences, and collaborate on solutions. These networks can take many forms, from local support groups to citizen-led organizations, all working towards the common goal of safeguarding our air and environment. By joining forces, communities can amplify their voices, drawing attention to the dangers of airborne chemicals and demanding change from authorities and policymakers.

But empowerment goes beyond simply raising awareness. It involves empowering individuals to take direct action and create tangible impact within their communities. This can be achieved through various means, such as organizing awareness campaigns, conducting scientific research, engaging with local leaders, and initiating grassroots initiatives. By working collectively, we can build momentum and make our voices heard on a wider scale, pressuring institutions to take action and enforce regulations that protect us from airborne chemical threats.

A crucial aspect of empowerment is the recognition that every individual possesses the ability to make a difference. By channeling our collective strength and determination, we can overcome the hurdles standing in our way. It is essential to break through the inertia of complacency and inspire individuals to step out of their comfort zones, realizing that their actions, no matter how small, hold the potential to drive significant change.

As we wrap up this chapter, we call upon every reader to become an agent of change within their own community. Let us empower ourselves, our families, and our neighbors to tackle the airborne chemical threat that looms over us. By taking a stand now, we can secure a healthier and safer future for generations to come.

However, our work does not end here. The second half of this chapter holds the key to unlocking the next level of our collective fight against airborne chemical threats. In the upcoming pages, we will explore groundbreaking initiatives, innovative technologies, and inspiring stories of individuals who have championed the cause. The battle against these invisible menaces is far from over, and our journey together is just beginning.

Remember, together we are strong, and together, we can make a lasting impact. Stay tuned for the second half of this chapter, where we will delve deeper into the strategies, successes, and challenges faced by communities worldwide as they strive to protect their air and future generations.

Across the globe, communities are rising up and taking a stand against the sinister threat of airborne chemicals. Empowered individuals refuse to let their health and the environment suffer in silence any longer. In this second half of the chapter, we delve deeper into the strategies, successes, and challenges faced by communities worldwide as they strive to protect their air and future generations.

One of the most successful strategies adopted by empowered communities is the organization of grassroots initiatives. These initiatives bring together passionate individuals who are committed to making a difference. From organizing public forums and town hall meetings to hosting workshops and educational campaigns, these initiatives aim to raise awareness and mobilize the community.

In some communities, concerned citizens have taken it upon themselves to conduct their own scientific research. Armed with knowledge and resources, they are shining a light on the specific dangers posed by airborne chemicals within their localities. These grassroots research projects not only provide critical evidence of the impact of these chemicals but also empower communities to demand accountability from authorities and policymakers.

However, every community faces its own unique set of challenges. Some may struggle with limited resources, while others may face opposition from powerful industrial interests. Overcoming these hurdles requires determination, collaboration, and the ability to build strong alliances. Communities have found success by forging partnerships with universities, non-profit organizations, and even government agencies that share their concerns.

In certain instances, communities have mobilized legal efforts to hold responsible parties accountable for the harm caused by airborne chemicals. Legal actions, ranging from class-action lawsuits to environmental impact assessments, serve as powerful tools to bring about change and ensure the protection of communities. By leveraging the legal system, communities are making it clear that the effects of airborne chemicals cannot go unnoticed.

The battle against airborne chemical threats has also seen innovative technologies emerge as a vital tool. Communities are harnessing the power of citizen science, developing sensor networks, and utilizing emerging technologies to monitor air quality and track the sources of contamination. This data-driven approach not only empowers residents with evidence but also strengthens their negotiating power when engaging with authorities and industries.

Communities worldwide have also achieved remarkable successes by actively engaging with local leaders, government officials, and policymakers. By organizing meetings, presenting scientific evidence, and advocating for stronger regulations, empowered individuals have made their voices heard and influenced policy changes at the local, regional, and even national levels. These successes prove that collective action and persistent advocacy can tip the scales in favor of a healthier and safer environment.

However, the journey to combat airborne chemical threats is not without its challenges. Pervasive misinformation, industry lobbying, and bureaucratic inertia continue to hinder progress. Overcoming these obstacles demands unwavering determination and a united front. Communities must remain vigilant, as the work to protect the air and future generations is an ongoing commitment.

As we conclude this chapter, it is important to remember that communities hold an immense power to effect change. Together, armed with knowledge, resilience, and unity, individuals can spearhead a movement that demands a healthier, safer, and more sustainable environment.

This chapter has provided a glimpse into the potential of communities to tackle airborne chemical threats head-on. By skipping no effort to educate, empower, and mobilize individuals, communities are becoming the driving force behind change. Remember, each one of us possesses the ability to take a stand and make a lasting impact.

The battle against airborne chemicals is far from over, but as we stand together, we hold the power to shape a future where the air we breathe is free of invisible menaces. Let us continue to unite, inspire, and advocate, ensuring that the legacy we leave behind is one of clean air, vibrant communities, and a brighter tomorrow for generations to come.

ABOUT THE AUTHOR

Frederic Spinola, the author of this book Chemtrails: The Silent Spread, is a passionate researcher and advocate for the pursuit of truth and knowledge. With a deep curiosity about the world and a commitment to unraveling complex phenomena, Spinola has dedicated significant time and effort to studying the controversial topic of chemtrails.

Drawing from a diverse range of disciplines, Spinola combines scientific rigor with a balanced approach, aiming to present a comprehensive exploration of chemtrails. Through extensive research and analysis, Spinola endeavors to provide readers with a nuanced understanding of the subject matter, delving into its historical context, scientific evidence, and societal implications.

While remaining open to different perspectives, Spinola approaches the topic with a critical mindset and seeks to foster informed discussions. By synthesizing information from various sources and engaging with experts in relevant fields, the author strives to present a well-rounded examination of the subject matter.

Spinola's dedication to uncovering the truth about chemtrails extends beyond the pages of this book. With a genuine desire to promote awareness and encourage further investigation, the author invites readers to explore the evidence, ask questions, and form their own conclusions.

Through the publication of this book, Frederic Spinola aspires to contribute to the ongoing conversation surrounding chemtrails and to empower readers with knowledge and a deeper understanding of this contentious topic.

www.ingramcontent.com/pod-product-compliance
Lightning Source LLC
Chambersburg PA
CBHW071046290526
45795CB00004B/1341